Advances in Nutrition and Top Sport

Medicine and Sport Science
Vol. 32

Series Editors
M. Hebbelinck, Brussels
R.J. Shephard, Toronto

Founder and Editor from 1969 to 1984
E. Jokl, Lexington, Ky.

Basel · München · Paris · London · New York · New Delhi · Bangkok · Singapore · Tokyo · Sydney

Advances in Nutrition and Top Sport

Volume Editor
F. Brouns, Maastricht

Associate Editors
W.H.M Saris, Maastricht
E.A. Newsholme, Oxford

36 figures and 18 tables, 1991

Basel · München · Paris · London · New York · New Delhi · Bangkok · Singapore · Tokyo · Sydney

Medicine and Sport Science

Published on behalf of the
International Council of Sport Science and Physical Education

---

Library of Congress Cataloging-in-Publication Data
  Advances in nutrition and top sport / volume editor, F. Brouns;
    associate editors, W.H.M. Saris, E.A. Newsholme.
    (Medicine and sport science; vol. 32)
  'Published on behalf of the Research Committee of the International Council of Sport Science
    and Physical Education'
  Includes bibliographical references and index.
  1. Sports -- Physiological aspects.   2. Athletes -- Nutrition.
  I. Brouns, F. (Fred)   II. Saris, Wilhelmus Hermanus Maria, 1949–.
  III. Newsholme, E.A.   IV. International Council of Sport Science and Physical Education.
  Research Committee.   V. Series.
  [DNLM: 1. Exercise.   2. Nutrition.   3. Sports.]
  ISBN 3–8055–5376–5 (alk. paper)

Drug Dosage
  The authors and the publisher have exerted every effort to ensure that drug selection and dosage set forth in this text are in accord with current recommendations and practice at the time of publication. However, in view of ongoing research, changes in government regulations, and the constant flow of information relating to drug therapy and drug reactions, the reader is urged to check the package insert for each drug for any change in indications and dosage and for added warnings and precautions. This is particularly important when the recommended agent is a new and/or infrequently employed drug.

All rights reserved.
  No part of this publication may be translated into other languages, reproduced or utilized in any form or by any means, electronic or mechanical, including photocopying, recording, microcopying, or by any information storage and retrieval system, without permission in writing from the publisher.

  © Copyright 1991 by S. Karger AG, P.O. Box, CH–4009 Basel (Switzerland)
  Printed in Switzerland on acid-free paper by Thür AG Offsetdruck, Pratteln
  ISBN 3–8055–5376–5

Contents

Preface	VII
Coyle, E.F. (Austin, Tex.): Carbohydrate Feedings: Effects on Metabolism, Performance and Recovery	1
Lemon, P.W.R. (Kent, Ohio): Does Exercise Alter Dietary Protein Requirements?	15
Anderson, R.A. (Beltsville, Md.): New Insights on the Trace Elements, Chromium, Copper and Zinc, and Exercise	38
Bendich, A. (Nutley, N.J.): Exercise and Free Radicals: Effects of Antioxidant Vitamins	59
Newsholme, E.A.; Parry-Billings, M.; McAndrew, N. (Oxford); Budgett, R. (Harrow): A Biochemical Mechanism to Explain Some Characteristics of Overtraining	79
Wurtman, R.J.; Lewis, M.C. (Cambridge, Mass.): Exercise, Plasma Composition, and Neurotransmission	94
Wagenmakers, A.J.M. (Maastricht): L-Carnitine Supplementation and Performance in Man	110
Rehrer, N.J. (Brussels): Aspects of Dehydration and Rehydration during Exercise	128
Maughan, R.J. (Aberdeen); Greenhaff, P.L. (Birmingham): High Intensity Exercise Performance and Acid-Base Balance: The Influence of Diet and Induced Metabolic Alkalosis	147
Brouns, F. (Maastricht): Gastrointestinal Symptoms in Athletes: Physiological and Nutritional Aspects	166
Saris, W.H.M. (Maastricht): Exercise, Nutrition and Weight Control	200
Subject Index	217

Preface

During the last decade the role of nutrition as an important factor in the preparation and performance of athletes has received renewed attention. A large number of studies have been performed in order to elucidate in particular the effect of exercise on metabolism and the effect of specific dietary measures on substrate supply and hormonal regulation.

The development of a 'fitness image' has led many people to start with some kind of endurance exercise and it was especially this development which has 'awakened' the sports-related industry. As a result, on the one hand, the scientific community has been supported by the food industry to perform studies in the field of nutrition and exercise, a collaboration which has been shown to be of utmost importance for scientific progress in this field. On the other hand, a large number of food products have overwhelmed the sports world, many of which are claimed to improve performance and health, without any scientific evidence.

In the volume presented here, the reader will find a number of chapters which stem from the special symposium 'Nutrition and Top Sport' which took place during the FIMS World Conference on Sports Medicine, May 1990, Amsterdam, The Netherlands. The goal of this symposium was to provide an update of classical sports nutritional topics, such as carbohydrate, protein, fluids and electrolytes and to introduce developing fields of scientific interest which particularly focus on health and well-being of athletes, such as gastrointestinal function, immune status and free radical pathology. In addition to these chapters, several authors have been invited to present a chapter on a specific sports nutritional topic, which has recently undergone new developments and/or has led to controversial interpretations and opinions.

As such, the chapters on trace elements, dietary influences on acid base balance, slimming and exercise and *L*-carnitine complete this book to a hopefully valuable source of information on the advancement of sports nutritional research. The symposium 'Nutrition and Top Sport' as well as the preparation of this volume were supported by an 'Isostar Science Grant' from Sandoz Nutrition, Bern, Switzerland.

Dr. *Fred Brouns*
Prof. *Wim H.M. Saris*
Prof. *Eric A. Newsholme*

Carbohydrate Feedings: Effects on Metabolism, Performance and Recovery

Edward F. Coyle

Human Performance Laboratory, Department of Kinesiology,
University of Texas at Austin, Tex., USA

Carbohydrate and fats are the predominant fuels oxidized by skeletal muscle to provide energy during prolonged exercise [1]. For more than a half century it has been repeatedly demonstrated that exercise of a moderate intensity (i.e., 60–80% maximal oxygen uptake, V_{O_2max}) cannot be maintained when carbohydrate stores within the body are not adequate [2, 3]. For reasons largely unknown, it is not possible to oxidize fat at rates greater than the energy requirements of exercise at approximately 50–60% V_{O_2max}, which corresponds to about 500 kcal/h in moderately trained people [4]. Therefore, it is necessary to oxidize carbohydrate during prolonged moderate intensity exercise. Muscle glycogen can be metabolized anaerobically at high rates (i.e., 1,500 kcal/h) for durations of approximately 1 min, whereas it can also supply approximately 700–800 kcal/h of aerobic energy for prolonged periods in moderately trained individuals. Another form of carbohydrate energy is blood glucose, which has the potential to be oxidized at rates of between 50 and 250 kcal/h, as discussed below.

Fat stores within the body are quite large, representing approximately 80,000 kcal even in lean young men. However, carbohydrate stores within the body are limited, with approximately 1,200 kcal available in the form of muscle glycogen and 400 kcal stored as liver glycogen and metabolized as blood glucose. Fatigue usually occurs when the muscle glycogen concentration in the exercising muscles reaches a critically low level [2]. Occasionally, fatigue may occur before muscle glycogen is depleted, due to the development of central nervous system hypoglycemia in sensitive individuals [3, 5]. Since fatigue during exercise often results from carbohydrate depletion, there has been much interest in determining the extent to which carbohydrate feedings before, during and after prolonged exercise can improve endurance performance and hasten recovery.

Carbohydrate Types and Metabolic Fat

Glucose availability, assessed by an increase in blood glucose following consumption of carbohydrate, is generally comparable for glucose, corn syrups, and pure starches, and somewhat less for sucrose (glucose + fructose). Vegetables, beans and other carbohydrates that contain mixtures of starches, fibers, and proteins can take longer to digest and, consequently, the appearance rate of glucose in blood is diminished. Fructose is converted to glucose relatively slowly by the liver, and the increase in blood glucose following fructose ingestion is usually very slight. Carbohydrates vary in sweetness characteristics. Fructose is most sweet, whereas sucrose and glucose are comparatively less sweet. Some corn syrups and maltodextrins taste 'starchy' or taste 'paste-like'. Concentrated sugar solutions ($>20\%$) can produce bloating, and ingestion of concentrated pure fructose is associated with gastrointestinal discomfort and diarrhea. Ingested carbohydrate is first transported by the bloodstream to the liver where it can be (1) converted to fat, (2) stored as glycogen, or (3) released into the bloodstream for transport to other tissues such as muscle.

Carbohydrate Needs of Athletes

Daily energy expenditure when training will obviously depend upon the distance covered and the intensity. Athletes who train intensely for 60–90 min expend 1,000–1,400 kcal in the process. These athletes generally must eat approximately 50 kcal of food/kg of body weight/day (23 kcal/lb/day), e.g. 3,500 kcal for a 70-kg (154 lb) athlete [6]. At least 50%, but ideally 60–70%, of the calories in endurance athletes' diets should be obtained from carbohydrate [7–9]. This amounts to approximately 500–600 g of carbohydrate (2,000–2,400 kcal/day). The remaining calories should be obtained from fat (20–30%) and protein (10–15%). Although most athletes recognize the importance of adequate dietary carbohydrate for high intensity training, their diets often contain less than 40% carbohydrate (350 g). As a result, they may feel chronically fatigued during periods of intense training.

Maximizing Muscle Glycogen prior to Competition

A few days prior to a prolonged and intense competitive event, athletes should regulate their diets and training in an attempt to maximize ('supercompensate' or 'load') muscle glycogen stores. High pre-exercise glycogen levels will allow athletes to exercise for longer periods by delaying fatigue [10, 11]. The most practical method of 'glycogen loading' involves altering training and diet 7 days. On days 7, 6, 5, and 4 before competition, one should train moderately hard (e.g., 1–2 h) and consume a moderately low carbohydrate diet (i.e., 350 g/day). This will make the muscle sufficiently carbohydrate deprived and ready to supercompensate, without making the person sick. During the 3 days prior to competition, training should be tapered (30–60 min/day) and a high carbohydrate diet consumed (i.e. 500–600 g/day). Such a regimen will increase muscle glycogen stores 20–40% or more above normal. This 'modified' glycogen loading regimen is as effective as the 'classic' regimen first proposed by Bergström and co-workers [2, 10], and more practical since it does not require athletes to attempt training while consuming a high fat diet.

The Precompetition Meal

High carbohydrate meals eaten within 6 h of competition may 'top-off' the glycogen stores in liver and muscle. The liver, which maintains blood glucose levels, relies upon frequent meals to sustain its small stores (80–100 g) of glycogen. Athletes who fast for 6–12 h prior to exercise and do not consume carbohydrate during exercise may experience a premature lowering of blood glucose during competition. Even after following a muscle glycogen loading regimen, it is wise to eat a low fat meal containing 75–150 g of carbohydrate during the 6 h prior to competition. If muscle glycogen stores are not already filled, a portion of the pre-exercise meal can be used to increase muscle glycogen prior to competition [12].

It has been suggested that athletes avoid high carbohydrate meals within 2 h of competition because carbohydrate may elevate blood insulin at the onset of exercise which could lead to a lowering of blood glucose during exercise [13]. It has been shown, however, that these responses are transient and probably will not impair performance [12, 14]. If carbohydrate is ingested during a warm-up, 'rebound hypoglycemia' does not occur and blood glucose is elevated during exercise [14]. In fact, several recent studies have found pre-exercise carbohydrate feedings to improve a person's ability to perform exercise at 70–80% V_{O_2max} after 2 h of exercise [15–18].

Carbohydrate Feeding during Prolonged Intense Exercise

After 1–3 h of continuous exercise at 60–80% V_{O_2max}, it is clear that athletes tire due to carbohydrate depletion. Carbohydrate feedings during exercise appear to delay fatigue by as much as 30–60 min [5, 19–22]. However, we have observed that this improvement in performance is not due to a sparing of muscle glycogen use during exercise. Instead, it appears that the exercising muscles rely mostly upon blood glucose for energy late in exercise [19].

The basis for this interpretation is shown in figure 1. During the second hour of prolonged exercise at 70–75% V_{O_2max}, well-trained cyclists display a progressive lowering of plasma glucose concentration (fig. 1a). When the subjects ingest a placebo solution, their fatigue after 3 h of cycling is preceded by a decline in the rate of carbohydrate oxidation (fig. 1b), and occurs at a time when muscle glycogen is low (fig. 1c) and the subjects are hypoglycemic (i.e., 2.5 mM). This agrees with the idea that fatigue is due to a lack of sufficient muscle glycogen and blood glucose. In contrast, when the subjects are fed carbohydrate throughout exercise (i.e., 70 g of maltodextrin at 20 min and 28 g every 30 min thereafter), plasma glucose concentration and the initial rates of carbohydrate oxidation are maintained. As a result, fatigue is delayed by 1 h (i.e., from 3 to 4 h). However, this improvement in performance was not due to a sparing of muscle glycogen use, as the rates of decline were identical during exercise with and without carbohydrate feedings (fig. 1c). Remarkably, the muscle glycogen use was minimal during the additional hour of exercise despite the fact that carbohydrate oxidation was maintained. This suggests that another source of carbohydrate was the predominant fuel during the latter stages of exercise, most likely blood glucose.

These concepts are summarized in figure 2. In well-trained cyclists, approximately 40–50% of energy for exercise at 70% V_{O_2max} is derived

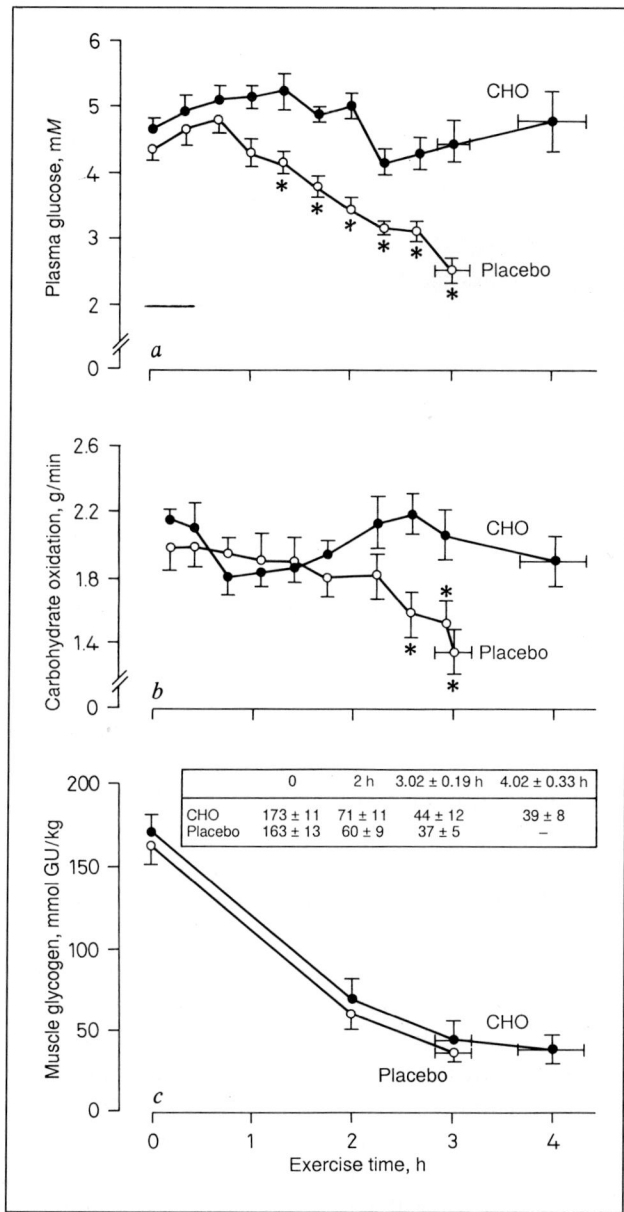

Fig. 1a-c. Responses when cycling at 74% V_{O_2max} with a placebo or when ingesting carbohydrate every 20 min (CHO). * Placebo different from carbohydrate; $p < 0.05$ [from 19].

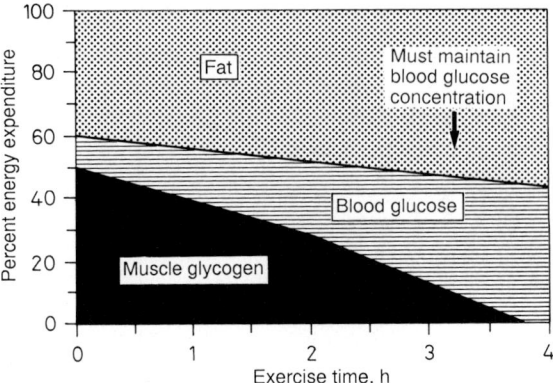

Fig. 2. Various sources of energy during prolonged exercise at 70% V_{O_2max}. Note that blood glucose becomes the predominant source of carbohydrate energy during the latter stages of exercise and thus it is important to maintain blood glucose concentration by eating carbohydrate.

from fat, whereas the remaining 50–60% is derived from carbohydrate. During the early portions of exercise, the majority of carbohydrate energy is derived from muscle glycogen. As exercise progresses, muscle glycogen is reduced and contributes less to the carbohydrate requirements of exercise. This reduced reliance upon muscle glycogen is balanced by an increased reliance upon blood glucose. After 3 h of exercise, the majority of carbohydrate energy appears to be derived from the metabolism of glucose, which is transported from the circulating blood into the exercising muscles.

Fatigue is not prevented by carbohydrate feeding, it simply is delayed. An obvious question regards the cause of fatigue when ingesting carbohydrate, since blood glucose and carbohydrate oxidation were high. It is possible that under these circumstances other factors besides carbohydrate depletion cause fatigue. It should be realized that during the final portions of exercise, when muscle glycogen is low and athletes rely heavily upon blood glucose for energy, their muscles feel tired and they must concentrate to maintain exercise at intensities that are ordinarily not stressful when muscle glycogen stores are filled. It is possible that the lowering of muscle glycogen plays a role in stimulating a high rate of muscle glucose uptake [20]. Additionally, late in exercise when muscle glycogen is low and blood glucose appears to be the major carbohydrate source, cyclists do not appear to be able to exercise more intensely than 75% V_{O_2max} for several minutes [22]. It is possible that this limitation reflects their inability to transport glucose from the blood and into the muscle fibers at the high rates required by more intense exercise.

Consuming Carbohydrate to Delay Fatigue

Consuming carbohydrate during prolonged continuous exercise will ensure that sufficient carbohydrate is available during the later stages of exercise. It has been demonstrated that fatigue can be reversed when carbohydrate supplementation is withheld until exhaustion [20]. However, the reversal of fatigue for 43 ± 5 min occurred only when athletes were infused intravenously with glucose at a rate of more than 1 g/min (or 250 kcal/h). This rate was required to meet the carbohydrate needs of exercising muscles. When these athletes drank 200 g of glucose in solution, they were unable to absorb this meal rapidly enough to maintain the energy needs of the exercising muscles as reflected by a decline in plasma glucose concentration (fig. 3). As a result they were able to maintain only an additional 26 ± 4 min of exercise. When a placebo was ingested, however, exercise could not be maintained for more that 10 ± 1 min. Therefore, people should ingest carbohydrate at regular intervals during exercise so that ample carbohydrate will be available when athletes will rely most heavily upon blood glucose for energy. A general recommendation is that 20–60 g of carbohydrate, in the form of maltodextrins, glucose or sucrose, be consumed every hour during prolonged exercise. Fructose feedings have not been observed to be effective for improving performance compared to glucose or sucrose probably because its conversion to and oxidation as glucose is not rapid enough to supply the carbohydrate energy requirements late in exercise [23, 24].

It also appears that carbohydrate feedings are beneficial during intermittent exercise [22, 23] as often is the case in sports such as soccer [25]. Carbohydrate feedings benefit activities which result in fatigue due to inadequate carbohydrate availability. It should be realized that muscle glycogen can become depleted after short duration high intensity exercise and future research should determine the extent to which carbohydrate feeding may benefit these activities. However, there appears to be no physiological need for carbohydrate supplementation during nonfatiguing exercise.

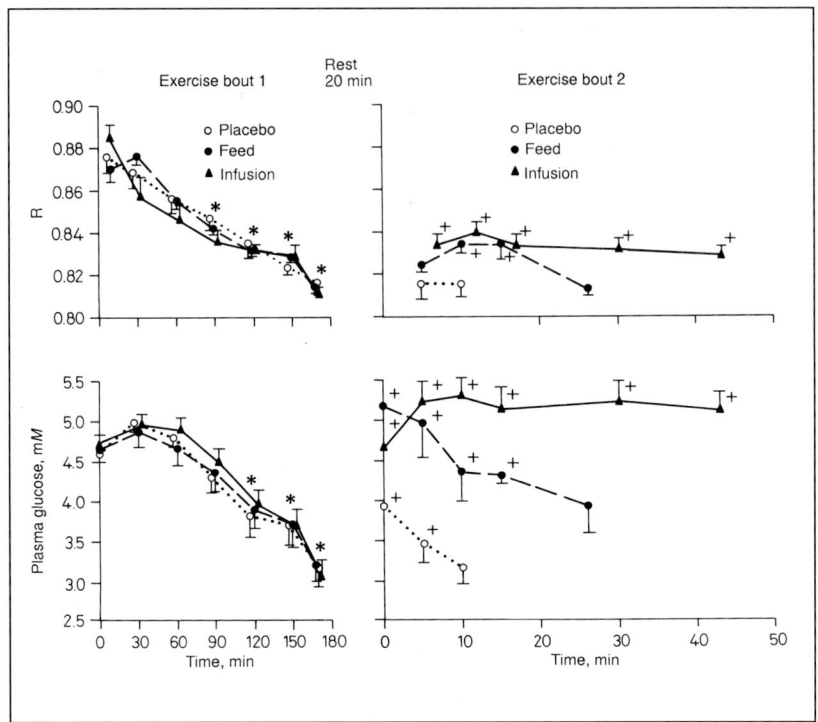

Fig. 3. During exercise bout 1, the subjects cycled at 74% V_{O_2max} until fatigued. After a 20-min rest, they continued to exercise (bout 2) with three different treatments. Placebo (o): after drinking flavored water and being infused with saline; infusion (▲) receiving intravenous glucose infusion at a rate which maintained plasma glucose at 5 mM; feeding (●): ingestion of 200 g of a 50% maltodextrin solution. * Significant decline during exercise bout 1; $p < 0.05$. + Values during exercise bout 2 which were significantly higher than at the end of exercise bout 1; $p < 0.05$.

Comparison of the Metabolic Responses to Low and High Intensity Exercise With and Without Carbohydrate Feeding

Carbohydrate feeding during prolonged exercise at ⩾70% V_{O_2max} does not cause hyperglycemia or hyperinsulinemia due to a catecholamine suppression of insulin release from the pancreas which is proportional to the intensity of exercise [5, 19]. Additionally, carbohydrate feeding during moderately intense exercise does not increase carbohydrate oxidation above fasting control levels during the first 2 h of exercise, but instead maintains carbohydrate oxidation at these levels late in exercise, which is the time when it would otherwise decline when fasted [5, 19].

In contrast, carbohydrate feeding during low intensity exercise (i.e., 30–50% V_{O_2max}) results in a 60% elevation of blood glucose and more than a twofold increase in plasma insulin concentration [26]. Also, the hyperinsulinemia is associated with a marked increase in leg glucose uptake and carbohydrate oxidation and a concomitant decline in fat oxidation [26]. Muscle glycogen during low intensity is not a major fuel and it appears that carbohydrate feeding shifts the extramuscular fuel source from free fatty acids to blood glucose.

Studies which have simultaneously determined both total and exogenous carbohydrate oxidation following carbohydrate feeding have observed exogenous utilization to increase more than total oxidation [27]. This indicates that endogenous carbohydrates are being spared. Although it is not possible to distinguish between a sparing of liver or muscle glycogen, it is likely that the ingested glucose replaces liver glycogen as the carbohydrate source which maintains blood glucose concentration. Therefore, it appears that the rationale for carbohydrate feeding during low intensity exercise would be to maintain liver glycogen. It is not clear, however, if hypoglycemia during prolonged low intensity exercise causes premature fatigue and therefore if carbohydrate feeding delays fatigue during mild long duration exercise.

There is little evidence to indicate that low intensity exercise is limited due to a lowering of blood glucose [28]. Although liver glycogen stores are lowered, gluconeogenesis can account for almost half of the glucose output by liver and therefore frank hypoglycemia is rare. After 13 h of walking at 33% V_{O_2max}, men still supply sufficient quantities of endogenous glucose to maintain the carbohydrate requirements of exercise [28]. Therefore, it remains to be determined if carbohydrate feeding during low intensity exercise, which seems to cause a shift from free fatty acid to blood glucose oxidation, improves endurance performance ability.

Extreme Endurance Training and Competition

The Tour de France represents the ultimate challenge. During this 22-day race, which covers 4,000 km including 30 mountain passes, cyclists eat an average of 5,800 kcal/day [7, 8]. Approximately 3,400 kcal (i.e., 60%

or 850 g) are from carbohydrates and approximately one-half of their caloric intake is consumed during the race, much in the form of 'sport drinks'.

In laboratory studies designed to simulate the Tour de France [8, 29], it was found that cyclists eating their normal diet could not eat enough carbohydrate both during and between exercise bouts to maintain energy balance. When given a fructose-maltodextrin energy drink supplement, energy balance was improved, but it still did not match expenditure. However, when consuming an athletic drink containing most of its carbohydrate in the from of maltodextrins, they were able to consume a sufficient amount of food to match their energy expenditure and they also maintain their body nitrogen balance [8]. When they were not able to balance their energy needs they relied more upon their bodies' protein stores for energy. It also appears that prolonged intense exercise when carbohydrate deprived increases accumulation of ammonia, which may also contribute to fatigue [30]. Although this is the extreme it also generally applies to athletes who train intensely day after day and who may not normally be ingesting enough carbohydrate in their diets or consuming it long enough before the next training session. This can lead to increased protein catabolism, which can be avoided provided that sufficient carbohydrates are ingested. It appears that athletes eating large amounts of carbohydrate will be able to better maintain caloric and nitrogen balance through maltrodextrin supplementation. Many athletes who appear 'stale' or 'overtrained', may actually be 'underfed' regarding carbohydrate.

Muscle Glycogen Resynthesis following Exercise

The rate of muscle glycogen resynthesis following exercise is obviously an important factor in recovery. In man, muscle glycogen is restored to normally highly levels at a rate of only about 3–7%/h (i.e., 5 mmol/kg muscle/h when attempting to increase muscle glycogen 100 mmol/kg). Therefore, under normal conditions, approximately 20 h are required to recover muscle glycogen stores and even a longer time will be necessary if diet is not optimal. The important dietary factors to consider are (1) time of food and drink after exercise, (2) the amount of carbohydrate eaten, and (3) the type of carbohydrate. It appears that the rate of muscle glycogen resynthesis is somewhat more rapid during the first 2 h following exercise compared to the 2- to 4-hour period (i.e., 7 vs. 4% respectively) [31].

Therefore, a recovering athlete should eat as soon after exercise as is practical. This will also provide more total time for resynthesis. When the results of two studies regarding optimal carbohydrate 'amount' are combined [32, 33], it appears that eating 25 g of carbohydrate every 2 h promotes resynthesis at 2%/h whereas eating anywhere from 50 to 225 g every 2 h causes glycogen to be resynthesized at 5–6%/h. Therefore, eating more than 50 g every 2 h does not seem beneficial. It is probably important that small frequent meals be eaten until a sufficient total amount of carbohydrate has been consumed (i.e., 600 g). Finally, glycogen is synthesized relatively slowly (i.e., 2%/h) when fructose is eaten compared to glucose or starch (i.e., 5–6%/h) [30]. Surprisingly, the rate of muscle glycogen resynthesis from sucrose was also 6%/h [32].

Muscle Glycogen Resynthesis during Exercise

The net decline in muscle glycogen concentration during exercise when ingesting carbohydrate will be determined by the balance between glycogenolysis and glycogen resynthesis. It does not appear that carbohydrate feedings alter the net rate of decline in muscle glycogen concentration during prolonged exercise maintained at a constant high intensity [19, 34]. However, it has been demonstrated in rats and in man that carbohydrate feedings given during low intensity exercise, which followed prolonged high intensity exercise, can promote glycogen resynthesis within nonactive muscle fibers with low glycogen concentration [35, 36]. However, this may not always occur [37]. Therefore, it is possible that carbohydrate feedings given during prolonged exercise which varies from high to low intensity, or which includes rest periods, may result in less of a net reduction in muscle glycogen concentration. Presumably this is due to glycogen resynthesis in fibers which are nonactive during the low intensity bouts of exercise. Along these lines, in the laboratory simulation of the Tour de France, which utilized exercise of intermittent intensity, it was observed that the decline in muscle glycogen was reduced by ingesting large amounts of carbohydrate during exercise [29]. Although it is not clear if this is due to decreased glycogenolysis or increased resynthesis during exercise, there seems to be good rationale for ingesting carbohydrate during intermittent exercise for the purpose of offsetting muscle glycogen depletion. This may be particularly important when racing repeatedly with little time for recovery and possibly inadequate time for full glycogen resynthesis.

Summary and Recommendations

Carbohydrate is the most important nutrient to cycling performance. The energy from carbohydrate can be released within exercising muscles up to three times as fast as energy from fat. However, carbohydrate stores in the body are limited; when depleted, athletes cannot exercise intensely and may experience fatigue. Muscle glycogen is maximized by following training hard during the week before competition and then progressively reducing the amount of daily exercise and eating a high carbohydrate diet (approximately 600 g/day) during the last 4 days before competition. Regular intense physical training requires consuming at least 50–60% of caloric intake as carbohydrate (approximately 400–600 g of carbohydrate daily). To maximize performance, it is often beneficial to consume 20–60 g of carbohydrate every hour during competition. Total replenishment of the body's carbohydrate stores requires a minimum of 20 h. For rapid muscle glycogen resynthesis, consume approximately 50–100 g of carbohydrate within 30 min after exercise followed by additional carbohydrate feedings every 2–4 h until a total of approximately 600 g have been eaten.

References

1. Krogh A, Lindhard J: Relative value of fat and carbohydrate as a source of muscular energy. Biochem J 1920;14:290–298.
2. Bergström J, Hultman E: The effect of exercise on muscle glycogen and electrolytes in normals. Scand J Clin Invest 1966;18:16–20.
3. Christensen EH, Hansen O: Untersuchungen über die Verbrennungsvorgänge bei langdauernder, schwerer Muskelarbeit. Skand Arch Physiol 1939;81:152–161.
4. Holloszy JO, Coyle EF: Adaptations of skeletal muscle to endurance exercise and their metabolic consequences. J Appl Physiol 1984;56:831–838.
5. Coyle EF, Hagberg JM, Hurley BF, et al: Carbohydrate feeding during prolonged strenuous exercise can delay fatigue. J Appl Physiol 1983;55:230–235.
6. Brotherhood JR: Nutrition and sports performance. Sports Med 1984;1:350–389.
7. Saris WHM, van Erp-Baart MA, Brouns F, et al: Study on food intake and energy expenditure during extreme sustained exercise: The Tour de France. Int J Sports Med 1989;10:S26–S31.
8. Brouns F, Saris WHM, Stroecken J, et al: Eating, drinking, and cycling. A controlled Tour de France simulation study. II. Effect of diet manipulation. Int J Sports Med 1989;10:S41–S48.
9. Costill DL, Sherman WM, Fink WJ, et al: The role of dietary carbohydrates in muscle glycogen resynthesis after strenuous running. Am J Clin Nutr 1981;34:1831–1836.

10 Bergström J, Hermansen L, Hultman E, et al: Diet, muscle glycogen, and physical performance. Acta Physiol Scand 1967;71:140–150.
11 Sherman WM: Carbohydrates, muscle glycogen, and muscle glycogen supercompensation; in Williams MH (ed): Ergogenic Aids in Sports. Champaign, Human Kinetics Publishers, 1983, pp 3–26.
12 Coyle EF, Coggan AR, Hemmert MK, et al: Substrate usage during prolonged exercise following a pre-exercise meal. J Appl Physiol 1985;59:429–433.
13 Costill DL, Coyle EF, Dalsky G, et al: Effects of elevated plasma FFA and insulin on muscle glycogen usage during exercise. J Appl Physiol 1977;43:695–699.
14 Brouns F, Rehrer NJ, Saris WHM, et al: Effect of carbohydrate intake during warming-up on the regulation of blood glucose during exercise. Int J Sports Med 1988;10: S68–S75.
15 Neufer PD, Costill DL, Flynn MG, et al: Improvements in exercise performance: effects of carbohydrate feedings and diet. J Appl Physiol 1987;62:983–988.
16 Gleeson M, Maughan RJ, Greenhaff PL: Comparison of the effects of preexercise feedings of glucose, glycerol and placebo on endurance and fuel homeostasis in man. Eur J Appl Physiol 1986;55:645–653.
17 Sherman WM, Brodowicz G, et al: Effects of 4 h preexercise carbohydrate feedings on cycling performance. Med Sci Sports Exerc 1989;21:598–604.
18 Hargreaves M, Costill DL, Fink WJ, et al: Effect of pre-exercise carbohydrate feedings on endurance cycling performance. Med Sci Sports Exerc 1987;19:33–36.
19 Coyle EF, Coggan AR, Hemmert MK, Ivy JL: Muscle glycogen utilization during prolonged strenuous exercise when fed carbohydrate. J Appl Physiol 1986;61:165–172.
20 Coggan AR, Coyle EF: Reversal of fatigue during prolonged exercise by carbohydrate infusion or ingestion. J Appl Physiol 1987;63:2388–2395.
21 Ivy JL, Miller W, Dover V, et al: Endurance improved by ingestion of a glucose polymer supplement. Med Sci Sports Exerc 1983;15:466–471.
22 Coggan AR, Coyle EF: Effect of carbohydrate feedings during high-intensity exercise. J Appl Physiol 1988;65:1703–1709.
23 Bjorkman O, Sahlin K, Hagenfeldt L, et al: Influence of glucose and fructose ingestion on the capacity for long-term exercise in well-trained men. Clin Physiol 1984;4: 483–494.
24 Murray R, Gregory LP, Seifert JG, et al: The effects of glucose, fructose and sucrose ingestion during exercise. Med Sci Sports Exerc 1989;21:275–282.
25 Foster C, Thompson N, Dean J, Krikendall D: Carbohydrate supplementation and performance in soccer players. Med Sci Sports Exerc 1986;18:S12.
26 Ahlborg G, Felig P: Influence of glucose ingestion on fuel-hormone response during prolonged exercise. J Appl Physiol 1976;41:683–688.
27 Pirnay F, Lacroix M, Mosora F, et al: Effect of glucose ingestion on energy substrate utilization during prolonged muscular exercise. Eur J Appl Physiol 1977;36:247–254.
28 Young DR, Pelligra R, Shapira J, et al: Glucose oxidation and replacement during prolonged exercise in man. J Appl Physiol 1967;23:734–741.
29 Brouns F, Saris WHM, Beckers E, et al: Metabolic changes induced by sustained exhaustive cycling and diet manipulation. Int J Sports Med 1989;10:S49–S62.

30 Brouns F, Beckers E, Wagenmakers AJM, Saris WHM: Ammonia accumulation during highly intensive longlasting cycling, individual observations. Int J Sports Med, in press.
31 Ivy JL, Katz AL, Cutler CL, Sherman WM, Coyle EF: Muscle glycogen synthesis after exercise: effect of time of carbohydrate ingestion. J Appl Physiol 1988;65: 1480–1485.
32 Blom PC, Hostmark AT, Vaage O, et al: Effect of different post-exercise sugar diets on the rate of muscle glycogen synthesis. Med Sci Sports Exerc 1987;19:491–496.
33 Ivy JL, Lee MC, Brozinick JT, et al: Muscle glycogen storage after different amounts of carbohydrate ingestion. J Appl Physiol 1988;65:2018–2023.
34 Hargreaves M, Briggs CA: Effect of carbohydrate ingestion on exercise metabolism. J Appl Physiol 1988;65:1553–1555.
35 Constable SH, Young JC, Higuchi J, et al: Glycogen resynthesis in leg muscles of rats during exercise. Am J Physiol 1984;247:R880–R883.
36 Kuipers H, Keizer HA, Brouns F, et al: Carbohydrate feeding and glycogen synthesis during exercise in man. Pflügers Arch (Eur J Physiol) 1987;410:652–656.
37 Kuipers H, Saris WHM, Brouns F, et al: Glycogen synthesis during exercise and rest with carbohydrate feeding in males and females. Int J Sports Med 1989;10:S63–S67.

Edward F. Coyle, PhD,
Human Performance Laboratory, Department of Kinesiology,
The University of Texas at Austin, Austin, TX 78712 (USA)

Does Exercise Alter Dietary Protein Requirements?

Peter W.R. Lemon[1]

Applied Physiology Research Laboratory, Kent State University, Kent, Ohio, USA

Introduction

Although the importance of protein in an athlete's diet has been debated for at least 150 years there is still no consensus on this issue. Athletes, especially strength athletes, generally consume vast amounts of protein [1–3] apparently believing that this practice increases muscle strength and size. In contrast, many nutritional/exercise scientists believe that the dietary protein needs of athletes are not appreciably greater than their sedentary counterparts [4, 5]. Recently, as a result of some new information this controversy has resurfaced in the scientific literature. The purpose of this paper is to review some of these new data and to discuss their relevance to various types of athletes.

Historical Overview

In order to put these recent data in their proper perspective, it is important to appreciate some history. In the mid-1800s the prevailing opinion was that protein was the primary fuel for working muscle [6]. Therefore, as might be expected athletes of the time consumed large amounts of dietary protein. In 1866, based on urinary nitrogen excretion

---

[1] Sincere appreciation is expressed to Ms. Janice Wilmoth for expert secretarial assistance.

measures (in order for protein to provide energy its nitrogen must be removed and subsequently excreted primarily in the urine), a classic paper (n = 2) was published which indicated that protein contributed about 6% of the fuel used during a 1,956 m climb in the Swiss Alps [7]. Although these data likely underestimated the actual protein use for several methodological reasons (the subjects consumed a protein-free diet before the climb, post-climb excretion measures were not made, and other routes of nitrogen excretion may have been substantial), this paper and several others [8] led to the perception that exercise does not increase one's need for dietary protein. Given this small contribution of protein relative to the belief at the time, this is understandable. However, based largely on these data, this belief has persisted throughout most of the 20th century [9]. This is somewhat surprising because Cathcart [8] in an extensive review of the literature prior to 1925 concluded 'the accumulated evidence seems to me to point in no unmistakable fashion to the opposite conclusion that muscle activity does increase, if only in small degree, the metabolism of protein'. It is important to realize that even a small percentage of a high exercise energy expenditure could produce an absolute increase in protein use and, therefore, increased requirements. Based on results from a number of separate experimental approaches, the conclusions of several recent investigators support Cathcart's conclusion [10–14]. Certainly no one has suggested that we return to the ideas of von Liebig on the significance of protein for exercise; however, it does appear that protein utilization, at least during endurance exercise, could be important enough to affect performance and may even increase an athlete's dietary protein needs. In addition, it is possible that protein intake in excess of recommended dietary allowances (RDA) could enhance gains in muscle mass and strength induced by strength exercise [15].

A review of the experiments on which the protein RDA was established [4, 5] indicates that little or no exercise was involved. Therefore, despite the traditional belief that exercise does not affect protein needs, it would appear that the effects of chronic intense exercise on protein requirements has not been adequately assessed. Furthermore, much of the available information regarding protein intake and muscle development is far from ideal as many studies have been limited by small sample sizes, inadequate control, and/or use of indirect measures. Perhaps future experiments will provide definitive data to help resolve these issues.

In summary, it would appear that over the past 150 years scientific opinion regarding the effect of exercise on dietary protein needs has changed several times. First, it was believed that protein was the major fuel for muscular work and as such was the most important component of an athlete's diet. When it became clear that this was in fact incorrect the observed small, but perhaps important, contribution of protein was all but overlooked. As a result, the prevailing opinion for most of the 20th century has been that exercise does not affect dietary protein needs. Why the observed protein contribution was considered unimportant is unclear. Perhaps it was simply an over-reaction to the realization that protein was not the major fuel source. In any event, recent experiments employing both the classic measures, as well as techniques not available to scientists at the turn of the century, have confirmed that protein does contribute to exercise fuel (perhaps 5–10%) at least with endurance exercise. In contrast, much less data are available concerning the role of dietary protein with strength exercise. Interestingly, over this entire time period, athletes, especially strength athletes, have continued to consume diets high in protein.

Due to available space it is not possible to review all studies that have investigated the effects of exercise on protein requirements. Rather, in an attempt to stimulate interest in a problem which has been largely overlooked for many years, I have limited this overview primarily to data from studies that challenge the belief that exercise does not alter protein requirements. Both extremes of the exercise intensity-duration continuum (endurance vs. strength exercise) are discussed. It is anticipated that these studies and those to be completed in the near future will provide the necessary data to answer the question posed with this review.

Effects of Endurance Exercise on Protein Requirements

Nitrogen Balance Experiments

Traditionally, nitrogen balance experiments (protein is approximately 16% nitrogen) are used to assess protein requirements. These experiments require quantification of all nitrogen consumed (food) as well as all nitrogen excreted (urine, sweat, faeces, etc.). Although this technique is extremely valuable with respect to overall protein needs, it is limited in that it provides no information regarding how protein metabolism is regulated. Further, because energy intake [16] and expenditure [17] affect nitrogen balance both must be carefully controlled. Obviously, these types of experiments are labor intensive, expensive, and difficult to control especially in human subjects. As a result, not many complete balance studies on individuals engaged in intense exercise have been published; however, several are of interest. For example, Brouns et al. [18, 19] completed an elaborate 7-day study where the effort of the Tour de France bicycle race was simulated in a respiration chamber using highly trained male cyclists (n = 13). The exercise consisted of 4.4 h/day at intensities ranging from 30 (warm-up) to 80% $\dot{V}_{O_2max}$ followed immediately by a 90% $\dot{V}_{O_2max}$ ride to exhaustion. Subjects consumed at least 1.4 g protein/kg·day^{-1} (175% protein RDA) and were in negative nitrogen balance. Due to the large energy expenditure, energy balance could only be maintained by ad libitum supplementation with a maltodextrin beverage. As expected, protein use was greater in the nonsupplemented subjects. The investigators recommended that athletes with these kinds of energy expenditures consume sufficient carbohydrate to maintain positive energy balance and dietary protein in the range of 1.5–1.8 g/kg·day^{-1} (188–225% protein RDA).

In an experiment involving less strenuous exercise, Gontzea et al. [20] studied nitrogen balance in young men who consumed 1.0 or 1.5 g protein/kg·day^{-1} (125–188% protein RDA) during 3 consecutive 4-day periods (sedentary pre-exercise; exercise 6, 20-min bouts/day, approximately 32–42 kJ/min; sedentary postexercise). While sedentary, both protein intakes resulted in positive nitrogen balance (more positive on the higher protein diet). However, during the exercise period daily nitrogen excretion increased and nitrogen balance became negative with the 1.0 g/kg·day^{-1} protein intake. Although the overall pattern was similar with the 1.5 g/kg·day^{-1} protein intake, nitrogen balance remained positive during the exercise period. With the cessation of daily exercise, nitrogen excretion decreased and nitrogen balance became positive in both groups (fig. 1). Because energy intake was sufficient to cover the added expenditure of the exercise, these data indicate that, at least at the beginning of an exercise program, a protein intake of 125% of the protein RDA is less than adequate even for this moderate exercise program. One could speculate that if this negative nitrogen balance were to persist over a long-term training program undesirable progressive losses in lean body mass would result. However, in a more prolonged exercise study (3 weeks) by the same group [21], it was found that this initial negative nitrogen balance was reduced over time indicating that during regular training some adaptation occurs leading to a conservation of body protein. These data suggest that dietary protein needs are greater during the initial period of training (perhaps 7–10 days) than they are later. In agreement with this possibility are the experiments of Yoshimura et al. [22] which indicate that increased dietary pro-

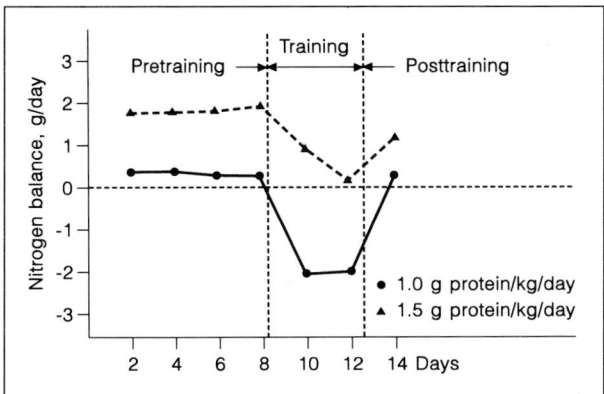

Fig. 1. Effect of endurance training on nitrogen balance when individuals consume 125 vs. 188% of the protein RDA. n = 30 per group. From Gontzea et al. [20].

tein is necessary to minimize the loss of blood proteins (sports anemia) that occurs during the first 10 days of an endurance training program. More recent studies [23, 24] also support the idea that the negative nitrogen balance observed with endurance exercise training becomes less pronounced over time. Because most exercise studies are relatively short term, these data raise the interesting question as to whether any increased protein need for endurance athletes is merely transient [25]. If so, one would expect the protein RDA (0.8 g/kg·day^{-1}) to be adequate for experienced endurance athletes.

To address this question it is necessary to have data from long-term exercise studies. These data are available from several experiments. Tarnopolsky et al. [26] studied nitrogen balance at 2 protein intakes in young men (trained endurance athletes, trained body builders, and sedentary controls). Throughout the study sufficient energy was consumed and regular activity (including training) was controlled. The protein intake necessary to obtain nitrogen balance (determined using linear regression) was 88% greater for the endurance athletes (1.37 g/kg·day^{-1}) when compared to the sedentary controls (0.73 g/kg·day^{-1}). In a similar study, Meredith et al. [27] studied nitrogen balance in endurance-trained men who were young (23–30 years) or middle aged (48–59 years). They found that protein needs were increased by 58% in these athletes and recommended (based on mean requirement plus twice the observed SD) a protein intake of

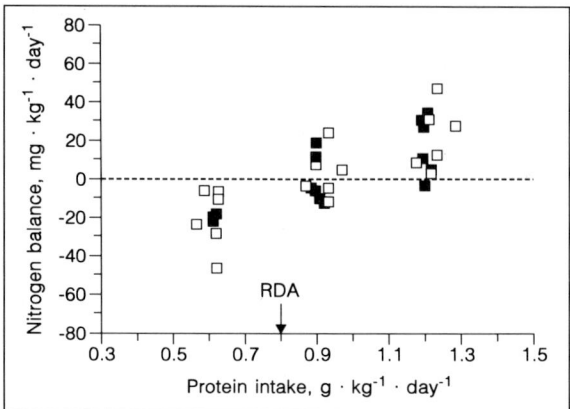

Fig. 2. Relationship between protein intake and nitrogen balance in young (□) and middle-aged (■) endurance-trained men. Data are for individual subjects. From Meredith et al. [27].

1.26 g/kg·day^{-1} (158% protein RDA) for endurance athletes (fig. 2). In a third study, Friedman and Lemon [28] essentially repeated the study described above by Gontzea et al. [20], except that endurance-trained men (24–29 years) were studied during their regular training (94 ± 21 km/week). While consuming 1.49 g/kg·day^{-1} (186% protein RDA) and adequate energy, nitrogen balance was positive. However, when 0.86 g/kg·day^{-1} (108% protein RDA) was consumed nitrogen balance was negative. Because the athletes in these 3 studies trained regularly for many years (2–40 years) these data indicate that increased dietary protein needs are not merely transient at the beginning of an endurance program.

Although the explanation for the observed differences in estimated protein requirements based on nitrogen balance data is not known, it may be due to differences in relative exercise intensity. The investigations that observed an increased requirement involved more intense exercise [26–28]. Further, the studies that observed a reduction in the negative nitrogen balance with time [21, 23, 24] used absolute workloads and, therefore, due to increased capacity with training may have actually studied lower relative exercise intensities as the studies progressed. Alternatively, the underlying mechanisms responsible for the increased protein requirement may differ comparing novice and experienced athletes [29]. Further study is necessary to assess these potential mechanisms. Together, these nitrogen balance data indicate that dietary protein needs are greater for athletes who engage in regular endurance exercise. This is especially important when energy expenditures are extremely high and/or energy intake is restricted (e.g. dieting). Precise recommendations must await further experiments but based on these data, protein needs of endurance athletes may be as high as 1.6 g/kg·day^{-1} (200% protein RDA).

Urea Studies

Urea is the major end product of protein metabolism in humans [30]. In the resting state urea is excreted primarily in the urine [31]; however, during exercise, especially if the exercise is prolonged or when glycogen availability is low, sweat losses become significant [12, 32, 33]. Several studies that have measured changes in serum urea concentration and/or urea excretion also suggest that endurance exercise could lead to increased protein needs.

Haralambie and Berg [34] studied serum urea and alpha amino nitrogen changes during competitive long distance (70–765 min) running, skiing, and/or walking races in 19- to 44-year-old men. Several observed changes appeared to be exercise duration related. After about 60–70 min of exercise, there was a significant increase in serum urea (greater than would occur due to exercise-induced decreased urea clearance) and a fall in alpha amino nitrogen. The investigators concluded that an increased breakdown of nitrogen-containing compounds occurred associated with exercise duration. Others have found similar results [35–37]. Because it is known that glycogen availability is reduced substantially by 60–70 min of intense exercise [38], Lemon and Mullin [12] manipulated glycogen stores (n = 6) prior to 60 min of cycle exercise at 61% $\dot{V}_{O_2max}$ and observed increased serum urea concentration and urea excretion in the glycogen-depleted men (27–30 years). These data indicate that glycogen availability may be the mechanism by which exercise duration influences dietary protein needs (i.e. when dietary carbohydrate is inadequate protein needs will be higher). To date, most of the available information is on male subjects. However, some data suggest that urea excretion with endurance exercise is greater during the luteal phase of the menstrual cycle [39] and/or may be absolutely less in female humans [40] (fig. 3) or rats [41]. Whether gender differences in protein utilization exist must await further investigation.

It should be noted that not all studies of urea excretion with endurance exercise have found increases; however, this may be because these studies are frequently shorter than nitrogen balance experiments. When postexer-

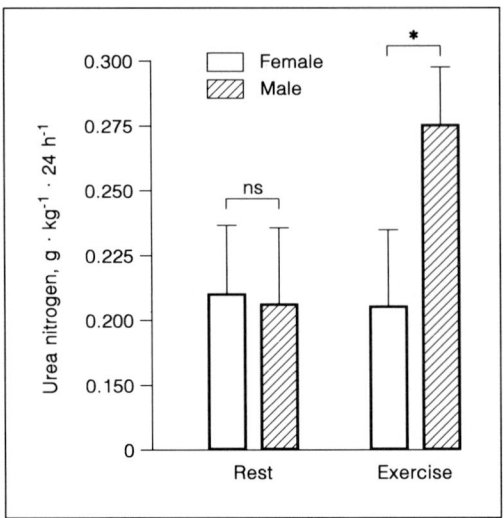

Fig. 3. Comparison of urea N excretion in men and women following a 15.5-km treadmill run at 65% $\dot{V}_{O_2max}$. n = 6 per group. *p < 0.05. From Tarnopolsky et al. [40].

cise collections continue for 1–2 days, increases are reported consistently [8, 35–37, 42–45]. This delayed urea excretion is likely caused by the effects of exercise dehydration on urea clearance [46] and can be prevented if adequate hydration is maintained [47]. Further, not all studies have quantified urea losses in exercise sweat and it is now apparent that this omission can result in an underestimate of protein use [12, 33, 48]. In summary although less complete, the observed urea data are consistent with the nitrogen balance results and strengthen the thesis that endurance exercise leads to increased dietary protein needs.

Amino Acid Oxidation Studies

Amino acids are the major constituents of proteins. In total, 20 different amino acids can be found in proteins but not every protein contains all 20 and, in addition, the relative proportions of each amino acid can vary widely in different proteins. Of these 20 amino acids, only 11 or perhaps 12 can be produced in the body. The remaining amino acids are called essential or indispensable because they must be obtained via diet. Therefore, the dietary requirement for protein is actually a requirement for select

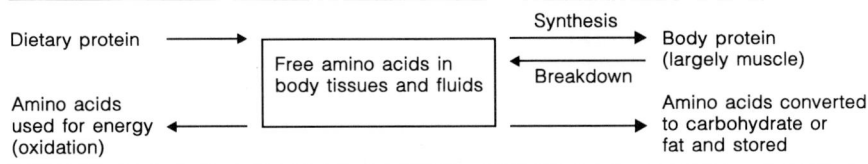

Fig. 4. Protein metabolism simplified.

amino acids and nitrogen necessary for the synthesis of the nonessential or dispensable amino acids.

During digestion, proteins are hydrolyzed and their constituent amino acids enter the free pool (fig. 4). Although this free pool is small (containing only a few percent of the body's total amino acids) it is central to protein/amino acid metabolism because all amino acids must pass through it. Absolute content of the free pool can be changed by altered amino acid input (diet and/or tissue protein breakdown) or output (protein synthesis, amino acid oxidation, amino acid conversion to carbohydrate or fat). Using stable or radiolabeled amino acids (injection or ingestion), it is possible to estimate rates of protein synthesis (incorporation of label) and amino acid oxidation (irreversible loss of label in breath) [49–51]. Unlike the nitrogen balance technique such measures allow conclusions about the mechanisms of protein balance.

In general, endurance exercise promotes decreases in protein synthesis [14, 52, 53] and perhaps increases in protein breakdown [54–56]. Either process by itself would increase the size of the free pool and the subsequent quantity of amino acids oxidized [57]. Although skeletal muscle has the capacity to oxidize at least 6 amino acids [58], the exercise studies to date have concentrated primarily on the amino acid leucine. These studies indicate that endurance exercise (in both humans and animals) increases substantially whole body leucine oxidation [13, 14, 55, 59–61]. The observed increases are severalfold higher in the human [62] and both high exercise intensity [13, 60, 61] and endurance training [11, 63] further increase the amount of leucine oxidized (fig. 5). The observed rates of leucine oxidation during exercise can approach daily leucine requirements [59]. Unless compensatory decreases occur during the rest of the day, it is likely that requirements for this amino acid are greater in individuals who regularly engage in endurance exercise. Similar determinations are needed for other amino acids to assess whether or not this pattern is unique.

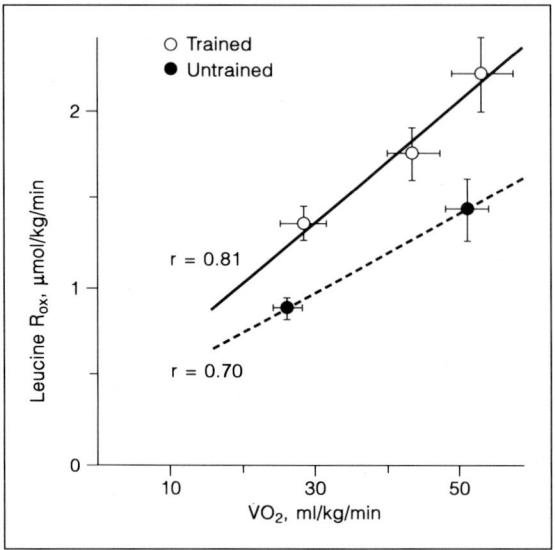

Fig. 5. Effect of exercise intensity and endurance exercise training on in vivo oxidation of the amino acid leucine. n = 16 (untrained) and 19 (trained) per group. From Henderson et al. [63].

It appears that activation of the limiting enzyme in the oxidation pathway of leucine (branched-chain keto acid dehydrogenase) by exercise intensity and/or duration is responsible for the observed increased leucine oxidation [64, 65]. Not only is the exercise activation significantly correlated with the increased exercise leucine oxidation but when exercise stops the decrease in leucine oxidation closely follows the decrease in branched chain keto acid dehydrogenase activity. In addition, Hood and Terjung [66] have demonstrated that activation of the branched chain keto acid dehydrogenase at rest (by inhibiting the kinase that inactivates it) produced increases in leucine oxidation similar to those obtained with endurance exercise. The mechanism underlying the endurance training effect on leucine oxidation is less clear because in situ muscle leucine oxidation rates are not increased with training [66]. These data indicate that, at least in the rat, changes in nonmuscle tissue are responsible for the observed increase in whole body leucine oxidation with training.

The observed increases in leucine oxidation with endurance exercise are consistent with increased dietary protein need. However, exercise does not increase the oxidation of all amino acids [51]. Therefore, rather than an increased protein need, endurance exercise may cause negative nitrogen balance because of an increased requirement for specific amino acids.

Tissue Degradation Markers

Several markers of muscle breakdown have been used to assess whether endurance exercise alters protein breakdown. These are usually quantified in the serum or urine (in vivo) or in the outflow from muscle (in situ or in vitro). Such markers include free tyrosine (or phenylalanine) and 3-methylhistidine. Tyrosine (or phenylalanine) can be used to estimate total protein degradation because it is not synthesized in the body (i.e. it is an indispensable amino acid) and it is not metabolized by muscle tissue. Therefore, an increased release from muscle indicates that tissue degradation must have occurred. However, in vivo this measure represents an underestimate because some tyrosine released from degraded protein could undergo protein synthesis. In vitro, this can be avoided by the use of an inhibitor of protein synthesis (e.g. cycloheximide). 3-Methylhistidine can be used to estimate myofibrillar protein breakdown because it is found exclusively in contractile proteins, and it cannot be re-incorporated into protein [67]. Therefore, when myofibrillar proteins are degraded 3-methylhistidine is released from skeletal muscle, and in vivo subsequently excreted in urine. Unfortunately, urinary excretion of 3-methylhistidine may be confounded by a significant contribution from smooth muscle degradation in the gut or skin [68, 69]; however, this seems to be less important in the human than the rat [70]. Therefore, this measure is generally considered to be a valid index of total body contractile protein breakdown. With exercise it is likely that any changes reflect primarily skeletal muscle degradation [71]. In vivo, 3-methylhistidine excretion is usually expressed as a ratio of urinary creatinine excretion in order to correct for differences in muscle mass among subjects as well as the effect of exercise on urine volume.

A large number of studies have investigated the urinary excretion of 3-methylhistidine in both rats and humans and found conflicting results [14, 36, 44, 73–75]. This may be due to the absence of adequate dietary control including the establishment of correct baseline values [76] and/or a possible biphasic response of 3-methylhistidine excretion (fig. 6) following exercise [71]. Perhaps the increased contractile protein degradation following endurance exercise is greater than the observed exercise reduction but this is not well documented.

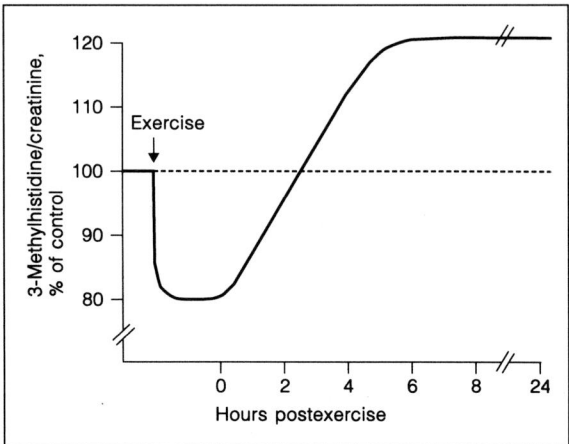

Fig. 6. Possible biphasic pattern of urinary 3-methylhistidine excretion during and following strenuous exercise. From Dohm et al. [71].

Increased serum tyrosine concentration has been observed during prolonged running or cycling in humans [35, 77]. Moreover, in vitro tyrosine release has been measured from muscles of rats following prolonged swimming or eccentric running [52, 78]. Because these tyrosine release data exceed the 3-methylhistidine results, Dohm et al. [71] have suggested that degradation of noncontractile proteins represents most of the observed protein content breakdown. Evidence for this suggestion is also provided from studies where decreases in both noncontractile muscle protein and liver protein content were measured following endurance exercise in rats [54]. Although these tissue degradation marker data are sketchy, they indicate an increased whole body protein breakdown with endurance exercise.

Effect of Strength Exercise on Protein Requirements

Nitrogen Balance Experiments
Several nitrogen balance experiments have been published indicating that the protein needs of strength athletes are greater than sedentary individuals; however, available data are quite variable. For example, Celejowa and Homa [79] reported that nitrogen balance was negative in 5 of 10 male

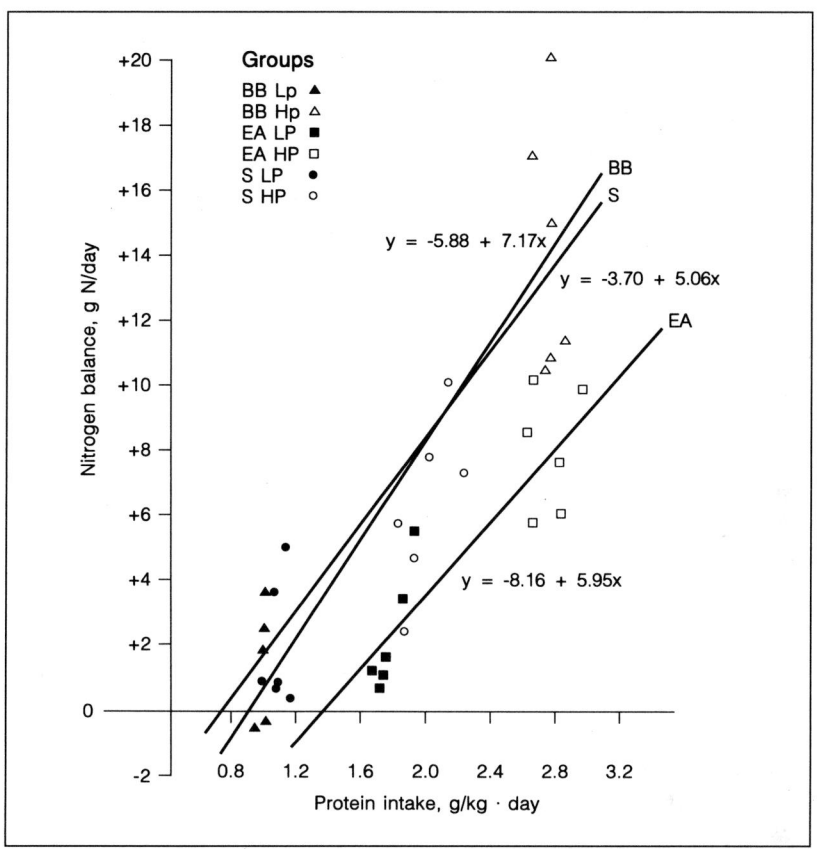

Fig. 7. Relationship between protein intake and nitrogen balance among bodybuilders, endurance athletes, and sedentary controls. n = 6 per group. From Tarnopolosky et al. [26].

weight lifters (20–35 years) even though they consumed about 2 g protein/ kg·day^{-1} (250% protein RDA). Although insufficient energy intake may have been responsible for the negative nitrogen balance in 1 subject, energy intake was sufficient in 4 of these 5 athletes. Therefore, these data suggest that the dietary protein need was substantially greater than the RDA for at least 40% of these lifters. Unfortunately, no information was provided about anabolic steroid use. In a more recent study which also used the nitrogen balance technique, Tarnopolsky et al. [26] found that the protein needs of young bodybuilders exceeded those of sedentary controls (fig. 7).

However, they estimated that this need could be met by 0.9 g/kg·day^{-1} (112% protein RDA) when energy intake was adequate. These subjects had not used anabolic steroids for at least 2 years. Other data [32] indicate that elite weight lifters can maintain positive nitrogen balance with protein intakes of 1.3–1.6 g/kg·day^{-1} (162–200% protein RDA) if energy intake is sufficient. Torun et al. [80] observed a negative nitrogen balance and a decreased cell mass (based on potassium-40 measures) with 6-weeks of strength training when protein intake was 100% of the protein RDA (n = 5). However, when protein intake was increased to 200% RDA (n = 2) nitrogen balance was positive and cell mass increased. Walberg et al. [81] have recently demonstrated the value of protein intakes in excess of the RDA for individuals engaged in strength exercise while on hypoenergetic diets. In this study while consuming 75.3 kJ/kg·day^{-1}, a protein intake of 1.6 g/kg (200% protein RDA) maintained positive nitrogen balance but 0.8 g/kg (100% protein RDA) resulted in a negative nitrogen balance. Although not directly relevant to the question of whether strength exercise elevates dietary protein needs, this study has considerable practical significance because of the large number of individuals who are dieting while engaged in strength training programs. Together, these nitrogen balance data indicate that protein needs of strength athletes are greater than sedentary individuals but apparently less than what many strength athletes consume.

These experimental results on strength athletes raise several questions. First, does high protein intake lead to progressively higher nitrogen balance or, alternatively, is the greater intake simply excreted? Second, if a greater positive nitrogen balance occurs with high protein intake is this beneficial for muscular development? Although neither question can be completely answered with the available literature some interesting data exist. Because adults are no longer growing it is generally believed that excess dietary nitrogen is simply excreted. However, in a well-controlled long-term study (105 days), Oddoye and Margen [82] demonstrated in 6 men (23–30 years) that a highly positive nitrogen balance persisted when protein intake was high (372% protein RDA). Given the powerful anabolic stimulus of strength exercise [83] it may be that the optimal stimulus for hypertrophy requires not just a positive, but a highly positive nitrogen balance [15]. If so, this could explain the disagreement over protein needs between strength athletes (whose opinions are the result of observed gains in muscle mass under uncontrolled conditions) and nutritional/exercise scientists (whose opinions are based primarily on results of short-term, well-controlled, nitrogen-balance experiments).

Consolazio et al. [48] studied 8 men (20–23 years) consuming either 1.4 g/kg·day^{-1} (175% protein RDA) or 2.8 g/kg·day^{-1} (350% protein RDA) over 40 days of rigorous exercise training (variety of activities totalling about 3,000 kJ/day). Adequate energy intake was provided and nitrogen balance measurements were made. Although both groups maintained positive nitrogen balance, the high protein group's nitrogen retention was much greater (32.4 vs. 7.1 g). Further, increases in lean body mass (by densitometry) were significantly greater in the high-protein group (3.28 vs. 1.21 kg). Marable et al. [84] also observed a greater nitrogen retention (estimated from dietary nitrogen minus urinary nitrogen) over 4 weeks of strength training when men (18–26 years) consumed 300% (n = 4) vs. 100% (n = 2) of the protein RDA. Dragan et al. [85] reported impressive gains in muscle strength (5%), muscle mass (6%), and lean body mass (5%, estimated from skinfold fat measures) over several months of strength training when world class weight lifters increased their dietary protein intake from 2.2 to 3.5 g/kg·day^{-1} (275–438% protein RDA). Unfortunately, these data may have been confounded by peaking for various championships. No information was given about anabolic steroid use. Finally, supplementation (23 g protein, 2,345 kJ/day) in a group of elderly men engaged in a 12-week leg strength training program resulted in greater gains in both thigh muscle and thigh fat mass (measured by computer-assisted tomography) when compared to nonsupplemented subjects [86]. However, no between group differences were observed in leg strength in this study.

Although some of these results support the athletes' belief that high protein diets are beneficial, it is premature to state that such diets actually enhance muscle gains in strength and size. Further, whether diets high in protein are harmful has not been adequately assessed. It has been suggested that high protein diets are associated with increased incidence of renal disease [87]; however, several populations of the world have protein intakes in the 250–300 g/day range (at least 450% protein RDA) and seem to tolerate these well [88]. Moreover, in an animal study [89] over 14 months, intake of diets containing 80% protein (most athletes consume <20% protein [1–3]) produced minimal negative effects.

Urea Studies

Most of the studies that have reported serum urea concentration or urinary urea excretion measures have involved endurance exercise. However, increased urea excretion has been observed in humans following strength exercise [44].

Amino Acid Oxidation Studies

Because amino acid oxidation increases with exercise intensity, it is possible that strength exercise could result in increased amino acid oxidation. However, due to the anaerobic nature of strength exercise such an increase would likely involve substantial increases in amino acid oxidation between exercise sets. Very little oxidation information is available but our preliminary data [90] indicate that changes in whole body leucine oxidation during and for 2 h following strength exercise (3 sets of 10 repetitions for 9 upper/lower body exercises at 70% 1 RM) are minimal.

Tissue Degradation Markers

Few tyrosine data are published with strength exercise perhaps because these data come primarily from animal experiments and for methodological reasons strength exercise has not been extensively investigated in humans. However, with other models of hypertrophy such as tenotomy or stretch tyrosine release from muscle has been shown to increase [91] even when increases in muscle protein synthesis occur [92].

Much more 3-methylhistidine data with strength exercise are available. Several studies used one bout of strength exercise [44, 93, 94] and have found conflicting results. As mentioned earlier, these data may be explained by the absence of adequate dietary control including establishment of correct baseline values [76] (fig. 8) or, because urine collection times varied, by a possible biphasic 3-methylhistidine exercise-recovery response [71]. Regardless, chronic strength exercise appears to produce a consistent elevation in urinary 3-methylhistidine excretion. Hickson and Hinkelmann [95] concluded that during the last 14 days of a 6-week strength training program (3 sets, 6 exercises, 6 days/week) urinary 3-methylhistidine excretion tended to increase. Frontera et al. [96] measured significant increases in 3-methylhistidine with 12 weeks (3 times/week) of strength training (knee flexion/extension) in older men (60–72 years). Pivarnik et al. [76] studied 11 men (24 ± 3.8 years) over 11 days of strength exercise (alternate upper/lower body exercises). 3-Methylhistidine excretion was significantly increased from days 3 through 11 when compared to a nonexercise baseline (established over 7 days prior to the study). Taken together it would appear that contractile protein degradation is increased, at least over the first 12 weeks of strength training. Because hypertrophy is the overall response to strength training [83], the rate of muscle protein synthesis must also increase, i.e. contractile protein turnover must be accelerated by strength training. Anabolic steroids in high

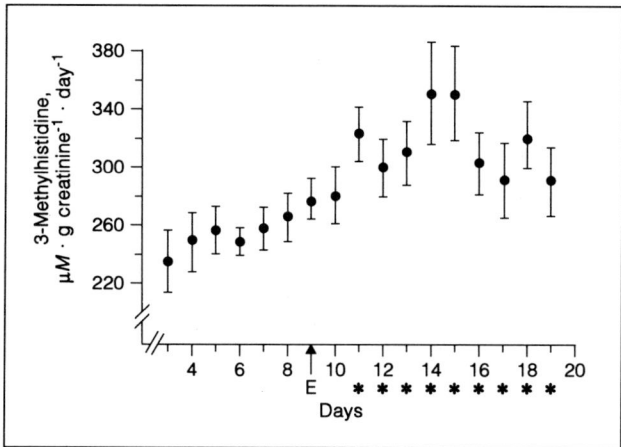

Fig. 8. Effect of repeated daily strength exercise on urinary 3-methylhistidine excretion. E indicates start of training. *Increase ($p < 0.001$) relative to pretraining values. n = 11. From Pivarnik et al. [76].

doses when combined with strength training appear to enhance this response resulting in greater increases in strength and size [97] but with a multitude of potential negative side effects [98].

Conclusions

For most of the 20th century nutritional/exercise scientists have believed that exercise does not affect dietary protein needs. However, a careful review of the data during the latter part of the 19th century indicates that the opposite may be true. Although protein provides a small percentage of total exercise energy, the absolute increase could affect protein needs. At least four separate lines of evidence collected over the past 20 years indicate that endurance exercise does increase protein need. Important contributing factors include exercise type, intensity and duration, degree of training, glycogen availability, and energy intake. Because these factors can vary independently, it is difficult to determine the exact dietary protein requirements but data suggest that needs may be as high as 1.6 g/kg·day^{-1} (200% protein RDA) for some endurance athletes. The situation is less clear with strength exercise but available evidence tends to suggest that increased protein intake may be beneficial for strength athletes as well. However, few data indicate that the quantity of protein consumed by many strength athletes is necessary. If energy intake is sufficient it would appear that a diet containing 12–15% of its energy from protein should be adequate for most athletes. Some athletes are also consuming large supplements of select free amino acids. For the most part the benefits/hazards of these supplements remain untested [99]. Further studies are needed to fine-tune dietary recommendations for athletes.

References

1. Katch VL, Katch FI, Moffatt, R, et al: Muscular development and lean body weight in body builders and weight lifters. Med Sci Sports Exerc 1980;12:340–344.
2. Short SH, Short WR: Four year study of university athletes' dietary intake. J Am Diet Assoc 1983;82:632–645.
3. Grandjean AC: Current nutritional beliefs and practices in athletes for weight/strength gains; in Garrett WE Jr, Malone TR (eds): Muscle Development: Nutritional Alternatives to Anabolic Steroids. Columbus, Ross Laboratories, 1988, pp 56–61.
4. Food and Nutrition Board: Recommended Dietary Allowances. Washington, National Academy Press, 1989, vol 10, pp 52–77.
5. Food and Agricultural Organization, World Health Organization, and United Nations University: Energy and Protein Requirements. Geneva, World Health Organization, 1985, Tech Rep Ser 724.
6. von Liebig J: Animal Chemistry or Organic Chemistry in Its Application to Physiology and Pathology (translated by W Gregory) London, Taylor & Walton, 1842, p 144.
7. Fick A, Wislicenus J: On the origin of muscular power. Phil Mag J Sci 1866;41:485–503.
8. Cathcart EP: Influence of muscle work on protein metabolism. Physiol Rev 1925;5:225–243.
9. Åstrand P-O, Rodahl K: Textbook of Work Physiology. New York, McGraw-Hill, 1977, p 487.
10. Poortmans JR: Effect of long lasting physical exercise and training on protein metabolism; in Howald H, Poortmans, JR (eds): Metabolic Adaptations to Prolonged Physical Exercise. Basel, Birkhäuser, 1975, pp 212–228.
11. Dohm GL, Hecker AL, Brown WE, et al: Adaptation of protein metabolism to endurance training. Increased amino acid oxidation in response to training. Biochem J 1977;164:705–708.
12. Lemon PWR, Mullin JP: Effect of initial muscle glycogen levels on protein catabolism during exercise. J Appl Physiol 1980;48:624–629.
13. White TP, Brooks GA: [U-^{14}C] glucose, -alanine, and leucine oxidation in rats at rest and two intensities of running. Am J Physiol 1981;241:E155–E165.

14 Rennie MJ, Edwards RHT, Krywawych S, et al: Effect of exercise on protein turnover in man. Clin Sci 1981;61:627–639.
15 Lemon PWR: Protein and exercise: Update 1987. Med Sci Sports Exerc 1987;19(5, suppl):S179–S190.
16 Munro HM: Carbohydrate and fat as factors in protein utilization and metabolism. Physiol Rev 1951;31:449–488.
17 Goranzon H, Forsum E: Effect of reduced energy intake versus increased physical activity on the outcome of nitrogen balance experiments in man. Am J Clin Nutr 1985;41:919–928.
18 Brouns F, Saris WHM, Stroecken, J, et al: Eating, drinking, and cycling. A controlled Tour de France simulation study, part I. Int J Sports Med 1989;10(suppl 1):S32–S40.
19 Brouns F, Saris WHM, Stroecken, J, et al: Eating, drinking, and cycling. A controlled Tour de France simulation study. II. Effect of diet manipulation. Int J Sports Med 1989;10(suppl 1):S41–S48.
20 Gontzea I, Sutzescu P, Dumitrache S: The influence of muscular activity on the nitrogen balance and on the need of man for proteins. Nutr Rep Int 1974;10:35–43.
21 Gontzea I, Sutzescu P, Dumitrache S: The influence to adaptation of physical effort on nitrogen balance in man. Nutr Rep Int 1975;11:231–234.
22 Yoshimura H, Inowe T, Yamada T, et al: Anemia during hard physical training (sports anemia) and its causal mechanism with special reference to protein nutrition. World Rev Nutr Diet 1980;35:1–86.
23 Butterfield GE, Calloway DH: Physical activity improves protein utilization in young men. Br J Nutr 1984;51:171–184.
24 Todd KS, Butterfield GE, Calloway DH: Nitrogen balance in men with adequate and deficient energy intake at three levels of work. J Nutr 1984;114:2107–2118.
25 Butterfield GE: Whole-body protein utilization in humans. Med Sci Sports Exerc 1987;19(5, suppl):S157–S165.
26 Tarnopolsky MA, MacDougall JD, Atkinson SA: Influence of protein intake and training status on nitrogen balance and lean body mass. J Appl Physiol 1988;64:187–193.
27 Meredith CN, Zackin MJ, Frontera WR, et al: Dietary protein requirements and body protein metabolism in endurance-trained men. J Appl Physiol 1989;66:2850–2856.
28 Friedman JE, Lemon PWR: Effect of chronic endurance exercise on retention of dietary protein. Int J Sports Med 1989;10:118–123.
29 Lemon PWR: Nutrition for muscular development of young athletes; in Gisolfi GV, Lamb DR (eds): Perspectives in Exercise Science and Sports Medicine: Youth, Exercise, and Sport. Indianapolis, Benchmark Press, 1989, vol 2, pp 369–400.
30 Newsholme EA, Leech AR: Biochemistry for the Medical Sciences. New York, Wiley, 1983, p 485.
31 Valtin H: Renal Function: Mechanisms Preserving Fluid and Solute Balance in Health. Boston, Little, Brown, 1973, p 136.
32 Laritcheva KA, Yalovaya NI, Shubin VI, et al: Study of energy expenditure and protein needs of top weight lifters; in Parizkova J, Rogozkin VA (eds): Nutrition, Physical Fitness, and Health. Baltimore, University Park Press, 1978, pp 155–163.

33 Cerny FJ: Protein metabolism during two hour ergometer exercise; in Howald H, Poortmans JR (eds): Metabolic Adaptations to Prolonged Physical Exercise. Basel, Birkhäuser, 1975, pp 232–237.
34 Haralambie G, Berg A: Serum urea and amino nitrogen changes with exercise duration. Eur J Appl Physiol 1976;36:39–48.
35 Refsum HE, Stromme SB: Urea and creatinine production and excretion in urine during and following prolonged heavy exercise. Scand J Clin Lab Invest 1974;33:247–254.
36 Decombaz J, Reinhardt P, Anantharaman, et al: Biochemical changes in a 100 km run: Free amino acids, urea and creatinine. Eur J Appl Physiol 1979;41:61–72.
37 Lemon PWR, Deutsch DT, Payne WR: Urea production during prolonged swimming. J Sports Sci 1989;7:241–246.
38 Gollnick PD, Piehl K, Karlsson J, et al: Glycogen patterns in human skeletal muscle fibers after varying types and intensities of exercise; in Howald H, Poortmans JR (eds): Metabolic Adaptations to Prolonged Physical Exercise. Basel, Birkhäuser, 1975, pp 416–421.
39 Lamont LS, Lemon PWR, Bruot BC: Menstrual cycle and exercise effects on protein catabolism. Med Sci Sports Exerc 1987;19:106–110.
40 Tarnopolsky LJ, MacDougall JD, Atkinson SA, et al: Gender differences in substrate for endurance exercise. J Appl Physiol 1990;68:302–308.
41 Dohm GL, Louis TM: Changes in androstenedione, testosterone and protein metabolism as a result of exercise. Proc Soc Exp Biol Med 1978;158:622–625.
42 Dunlop JC, Paton DN, Stockman R, et al: On the influence of muscular exercise, sweating, and massage, on the metabolism. J Physiol 1897;22:68–91.
43 Lemon PWR, Dolny DG, Sherman BA: Effect of intense prolonged running on protein catabolism; in Knuttgen HG, Vogel JA, Poortmans JR (eds): Biochemistry of Exercise. Boston, Human Kinetics, 1983, pp 367–372.
44 Dohm GL, Williams RT, Kasperek GJ, et al: Increased excretion of urea and N^+-methylhistidine by rats and humans after a bout of exercise. J Appl Physiol 1982;52:27–33.
45 Dolny DG, Lemon PWR: Effect of ambient temperature on protein breakdown during prolonged exercise. J Appl Physiol 1988;64:550–555.
46 Austin JH, Stillman E, Van Slyke DD: Factors governing the excretion of urea. J Biol Chem 1921;46:91–112.
47 Dolan PL, Hackney AC, Lemon PWR: Importance of hyration on protein catabolism estimates made from urinary urea measures. Med Sci Sports Exerc 1987;19(2, suppl):S33.
48 Consolazio CF, Johnson HL, Nelson RA, et al: Protein metabolism during intensive physical training in the young adult. Am J Clin Nutr 1975;28:29–35.
49 Waterlow, JC; Garlick PJ, Millward DJ: Protein Turnover in Mammalian Tissues and in the Whole Body. Amsterdam, Elsevier/North Holland, 1978.
50 Matthews DE, Bier DM: Stable isotope methods for nutritional investigation. Ann Rev Nutr 1983;3:304–339.
51 Wolfe RR: Tracers in Metabolic Research: Radioisotope and Stable Isotope/Mass Spectrometry Methods. New York, Liss, 1984.
52 Dohm GL, Kasperek GJ, Tapscott EB, et al: Effect of exercise on synthesis and degradation of muscle protein. Biochem J 1980;188:255–262.

53 Booth FW, Watson PA: Control of adaptations in protein levels in response to exercise. Fed Proc 1985;44:2293–2300.
54 Dohm GL, Puente FR, Smith CP, et al: Changes in tissue protein levels as a result of endurance exercise. Life Sci 1978;23:845–850.
55 Lemon PWR, Benevenga NJ, Mullin JP, et al: Effect of daily exercise and food intake on leucine oxidation. Biochem Med 1985;33:67–76.
56 Ji LL, Stratman FW, Lardy HA: Enzymatic down regulation with exercise in skeletal muscle. Arch Biochem Biophys 1988;263:137–149.
57 Harper AE: Control mechanisms in amino acid metabolism; in Sink JD (ed): Control of Metabolism. University Park, Penn State University Press, 1974, pp 49–74.
58 Goldberg AL, Odessey R: Oxidation of amino acids by diaphragms from fed and fasted rats. Am J Physiol 1972;223:1384–1391.
59 Evans WJ, Fisher EC, Hoerr RA, et al: Protein metabolism and endurance exercise. Phys Sportsmed 1983;11:63–72.
60 Babij P, Matthews SM, Rennie MF: Changes in blood ammonia, lactate and amino acids in relation to workload during bicycle ergometer exercise in man. Eur J Appl Physiol 1983;50:405–411.
61 Lemon PWR, Nagle FJ, Mullin JP, et al: In vivo leucine oxidation at rest and during two intensities of exercise. J Appl Physiol 1982;53:947–954.
62 Hood DA, Terjung RL: Amino acid metabolism during exercise and following endurance training. Sports Med 1990;9:23–35.
63 Henderson SA, Black AL, Brooks GA: Leucine turnover and oxidation in trained rats during exercise. Am J Physiol 1985;249:E137–E144.
64 Kasperek GJ, Dohm GL, Snider RD: Activation of branched-chain keto acid dehydrogenase by exercise. Am J Physiol 1985;248:E166–E177.
65 Kasperek GJ, Snider RD: Effect of exercise intensity and starvation on activation of branched-chain keto acid dehydrogenase by exercise. Am J Physiol 1987;252:E33–E37.
66 Hood DA, Terjung RL: Effect of endurance training on leucine metabolism in perfused rat skeletal muscle. Am J Physiol 1987;253:E648–E656.
67 Young VR, Munro HN: N^+-methylhistidine (3-methylhistidine) and muscle protein turnover: An overview. Fed Proc 1978;37:2291–2300.
68 Rennie MJ, Millward DJ: 3-Methylhistidine excretion and the urinary 3-methylhistidine/creatinine ratio are poor indicators of skeletal muscle protein breakdown. Clin Sci 1983;65:217–225.
69 Wassner SJ, Li JB: N^+-methylhistidine release: Contributions of rat skeletal muscle GI (gastrointestinal) tract, and skin. Am J Physiol 1982;243:E293–E247.
70 Afting E-G, Berhart W, Janzen RWC, et al: Quantitative importance of nonskeletal muscle N^+-methylhistidine and creatinine in human urine. Biochem J 1981;200:449–452.
71 Dohm GL, Tapscott EB, Kasperek GJ: Protein degradation during endurance exercise and recovery. Med Sci Sports Exerc 1987;19(5, suppl):S166–S171.
72 Frish RE, Hall GM, Aoki TT, et al: Metabolic, endocrine, and reproductive changes of a woman channel swimmer. Metabolism 1984;33:1106–1111.
73 Kasperek GJ, Snider RD: Increased protein degradation after eccentric exercise. Eur J Appl Physiol 1985;54:30–34.

74 Evans, WJ, Meredith CN, Cannon JG, et al: Metabolic changes following eccentric exercise in trained and untrained men. J Appl Physiol 1986;61:1864–1868.
75 Mussini E, Columbo L, DePonte G, et al: Effect of swimming on protein degradation: 3-methylhistidine and creatinine excretion. Biochem Med 1985;34:373–375.
76 Pivarnik JM, Hickson JF, Wolinsky I: Urinary 3-methylhistidine excretion increases with repeated weight training exercise. Med Sci Sports Exerc 1989;21:283–287.
77 Bergström J, Furst P, Noree LO: Intracellular free amino acid concentration in human muscle tissue. J Appl Physiol 1974;36:693–697.
78 Kasperek GJ, Snider RD: The effect of exercise on protein turnover in isolated soleus and extensor digitorum longus muscle. Experientia 1985;41:1399–1400.
79 Celejowa I, Homa M: Food intake, nitrogen and energy balance in Polish weight lifters, during a training camp. Nutr Metabol 1970;12:259–274.
80 Torun B, Scrimshaw NS, Young VR: Effect of isometric exercises on body potassium and dietary protein requirements of young men. Am J Clin Nutr 1977;30:1983–1993.
81 Walberg JL, Leidy MK, Sturgill DJ, et al: Macronutrient content of a hypoenergy diet affects nitrogen retention and muscle function in weight lifters. Int J Sports Med 1988;9:261–266.
82 Oddoye EB, Margen S: Nitrogen balance studies in humans: Long-term effect of high nitrogen intake on nitrogen accretion. J Nutr 1979;109:363–377.
83 Goldberg AL, Etlinger JD, Goldsprink DF, et al: Mechanisms of work-induced hypertrophy of skeletal muscle. Med Sci Sports 1975;7:248–261.
84 Marable NL, Hickson JF, Korslund MK, et al: Urinary nitrogen excretion as influenced by a muscle-building exercice program and protein intake variation. Nutr Rep Int 1979;19:795–805.
85 Dragan GI, Vasiliu A, Georgescu E: Effect of increased supply of protein on elite weight-lifters; in Galesloot TE, Tinbergen BJ (ed): Milk Proteins '84. Wageningen, Pudoc, 1985, pp 99–103.
86 Meredith CN: Protein needs and protein supplements in strength-trained men; in Garrett WE Jr, Malone TR (eds): Muscle Development: Nutritional Alternatives to Anabolic Steroids. Columbus, Ross Laboratories, 1988, pp 68–72.
87 Brenner BM, Meyer TW, Hostetter TH: Dietary protein intake and the progressive nature of kidney disease: The role of hemodynamically mediated glomerular injury in the pathogenesis of progressive glomerular sclerosis in aging, renal ablation and intrinsic renal disease. N Engl J Med 1982;307:652–659.
88 Durnin JVGA: Protein requirements and physical activity; in Parizkova J, Rogozkin VA (eds): Nutrition, Physical Fitness, and Health. Baltimore, University Park Press, 1978, pp 53–60.
89 Zaragoza R, Renau-Piqueras J, Portoles M, et al: Rats fed prolonged high protein diets show an increase in nitrogen metabolism and liver megamitochondria. Arch Biochem Biophys 1987;258:426–435.
90 Tarnopolsky MA, Atkinson SA, MacDougall JD, et al: Leucine turnover during and after weightlifting in young men. Med Sci Sports Exerc 1990;22(2, suppl):in press.
91 Baracos VE, Goldberg AL: Maintenance of normal strength improves protein balance and energy status in isolated rat skeletal muscles. Am J Physiol 1986;251:C588–C596.

92 Augert G, Monier S, Le Marchand-Brustel Y: Effect of exercise on protein turnover in muscles of lean and obese mice. Diabetologia 1986;29:248–253.
93 Hickson JF, Wolinsky I, Rodriguez, et al: Failure of weight training to affect urinary indices of protein metabolism in men. Med Sci Sports Exerc 1986;18:563–567.
94 Horswill CA, Layman DK; Boileau RA, et al: Excretion of 3-methylhistidine and hydroxyproline following acute weight-training exercise. Int J Sports Med 1988;9:245–248.
95 Hickson JF, Hinkelmann K: Exercise and protein intake effects on urinary 3-methylhistidine excretion. Am J Clin Nutr 1985;41:246–253.
96 Frontera WR, Meredith CN, O'Reilly KP, et al: Strength conditioning in older men: Skeletal muscle hypertrophy and improved function. J Appl Physiol 1988;64:1038–1044.
97 Hervey GR, Hutchinson I, Knibbs AV, et al: Anabolic effects of methandienone in men undergoing athletic training. Lancet 1976;ii:699–702.
98 Ryan AJ: Anabolic steroids are fool's gold. Fed Proc 1981;40:2682–2688.
99 Lemon PWR, Chaney MM: Physiologic effects of amino acid supplementation; in Garrett WE Jr, Malone TR (eds): Muscle Development: Nutritional Alternatives to Anabolic Steroids. Columbus, Ross Laboratories, 1988, pp 62–67.

Dr. Peter W.R. Lemon, Applied Physiology Research Laboratory,
Kent State University, Kent, OH 44242 (USA)

New Insights on the Trace Elements, Chromium, Copper and Zinc, and Exercise

Richard A. Anderson

Vitamin and Mineral Nutrition Laboratory, Beltsville Human Nutrition Research Center, US Department of Agriculture, ARS, Beltsville, Md., USA

The trace elements, chromium, copper and zinc, are involved individually or collectively in essentially all phases of energy production and utilization including breakdown of by-products of energy production, e.g., lactic dehydrogenase, a zinc-dependent enzyme, plays a critical role in the removal of lactic acid from exercising muscles. These trace elements are often overlooked when attempts are made to improve exercise performance through improved nutrition. If trace elements are supplemented, unbalanced nutrient supplements are usually employed. Nutrition studies in athletes usually concentrate on the macroelements in the diet such as type and amount of protein, fat or carbohydrate. Utilization, function and storage of these macroelements are in fact controlled by trace elements.

There is a complex series of interactions between dietary components leading to a delicate balance that is often regulated by trace elements such as chromium, copper and zinc. Therefore, a knowledge and understanding of the role of trace elements in energy production and utilization may be critical to attaining optimal health and performance.

This communication is an attempt to explain how exercise effects trace element function and what effects insufficient dietary intake of trace elements could have not only on exercise performance but also on overall health. This review updates and expands previous reviews on trace elements and exercise [1–3].

Chromium

Chromium is involved in normal carbohydrate and lipid metabolism where it functions primarily by potentiating insulin activity. In the presence of sufficient amounts of chromium in a usable form, much lower amounts of insulin are required. Chromium also may be involved in maintaining the structural integrity of nucleic acids.

Most individuals and animals can convert inorganic chromium to a usable form. However, genetically diabetic mice and possibly brittle diabetics and/or people with advanced stages of maturity-onset diabetes lose the ability to convert chromium to a usable form and are therefore dependent upon preformed physiologically active forms of chromium [4].

Stresses including high sugar diets, physical trauma, infection, certain diseases and strenuous exercise exacerbate signs and symptoms of marginal chromium deficiency [5]. The normal suboptimal intake of chromium, increased losses associated with strenuous exercise plus the added drain on chromium reserves associated with intakes of large amounts of simple sugars, suggest that chromium status of individuals who exercise strenuously and consume high simple sugar diets may be compromised.

Signs of Deficiency, Dietary Chromium Intake and Requirement

Signs of marginal chromium deficiency that are widespread in most population groups include impaired glucose tolerance, elevated insulin, glycosuria, decreased insulin receptor number, elevated cholesterol and triglycerides, decreased HDL-cholesterol and hypoglycemia [5]. Decreased fertility and sperm count is an early sign of marginal chromium deficiency in rats [6] but has not been studied in humans.

In 1980, the recommended or suggested safe and adequate intake for chromium for adults was established at 50–200 µg/day [7]. However, chromium intake in the US and other developed countries is usually 50–60% of the minimum suggested safe and adequate intake. Anderson and Kozlovsky [8] reported a mean daily chromium intake of 25 ± 1 µg for females and 33 ± 3 µg for males. Similar results have been reported from England [9], and Finland [10]. Slightly higher intakes were reported for Canadian subjects [11]. Absorption of chromium is inversely related to dietary intake at normal chromium intakes of less than 40 µg/day. Normal chromium absorption usually varies from 0.5 to 4% [8].

Chromium requirement appears to be related to degree of impaired glucose tolerance. Control normal subjects appear to have the lowest chromium requirement followed by subjects with marginally impaired glucose tolerance and hypoglycemics and a higher requirement for maturity-onset diabetics [12]. Subjects with either hyper- or hypoglycemic values often respond to supplemental chromium. Normal subjects with good glucose tolerance that do not show signs of chromium deficiency, would not be expected to and do not respond to supplemental chromium. Chromium functions as a nutrient and not a therapeutic agent.

Good sources of chromium include bran cereals, broccoli, some processed meats, oysters, mushrooms and some brands of beer and wine [2]. The chromium content of food varies widely not only among different types of foods but also among different lots or sources of the same foods. Substantial amounts of chromium appear to be introduced from exogenous sources during growing, transport, processing and in preparation of the final food product. However, this exogenous chromium appears to be absorbed and utilized [13].

Not only is the dietary intake of chromium important, but other foods that alter chromium metabolism and losses may be crucial. Simple sugars enhance chromium losses [14]. Foods high in simple sugars, e.g., pastries, soft drinks and refined baked goods, enhance chromium losses and are also usually low in chromium. Therefore, these foods that may be consumed during periods of increased carbohydrate intake may lead to depleted chromium stores due to low intake and increased losses.

Toxicity of chromium in the plus three valance state, the form found in foods and supplements, is rare due to low chromium absorption. Signs of chromium toxicity are almost exclusively restricted to chromium in the plus six valence state. Chromium (VI) is usually encountered in industrial settings and usually does not pose a health problem for the overwhelming majority of the population. Chromium does interact with other nutrients, especially iron and zinc, and excessive intakes of chromium should be avoided [15].

Exercise Effects on Blood Chromium

Blood chromium is not a meaningful indicator of chromium status since it is not in equilibrium with body stores. Blood chromium is also difficult to measure and often subject to contamination [15]. Urinary losses are a cumulative total of small transitory changes in the blood and appear to be a more accurate reflection of changes in chromium metabolism.

Immediately following a 10-km run, serum chromium was elevated approximately 50% above preexercise levels and remained elevated 2 h following exercise [16]. Basal serum chromium of trained and untrained subjects was similar [17]. Serum chromium of trained subjects increased significantly 2 h following treadmill exercise to exhaustion (30 s exercise, 30 s rest). Serum chromium of untrained subjects was similar to preexercise levels 2 h postexercise. These exercise-induced changes in blood chromium were also reflected by increased urinary chromium levels.

Chromium Losses and Exercise

The urine is the primary route of excretion for absorbed chromium. Once chromium is mobilized it does not appear to be reabsorbed but is excreted via the urine [15]. Therefore, urinary chromium losses are a meaningful indicator of chromium utilization.

Exercise leads to severalfold increases in urinary chromium concentration. Mean urinary chromium concentration increased nearly 5-fold 2 h following a strenuous 10-km run [18]. Similar results were obtained when chromium losses were expressed per milligram of creatinine to correct for changes in urine volume. Total daily urinary chromium excretion was roughly 2 times higher the day of running compared with a nonrun day. This is not a nonspecific exercise effect on metal losses since daily urinary excretion of sodium, potassium and calcium were unchanged due to exercise.

In the study of Anderson et al. [18], data were not corrected for degree of exercise training and effects due to altered dietary intake of chromium and other nutrients were not constant. In a follow-up study [17], 8 physically trained and 5 sedentary subjects were placed on a constant diet. Subjects with V_{O_2max} values in the good or above range based on their age were designated as trained and subjects with V_{O_2max} values in the average or below were designated as untrained [19]. Basal urinary chromium excretion of the trained subjects was significantly lower than that of the sedentary control subjects, 0.09 ± 0.01 and 0.21 ± 0.03 µg/day, respectively. Basal chromium losses of the trained subjects may be lower than those of sedentary subjects due to partial depletion of body stores resulting from repeated bouts of strenuous exercise that cause increased chromium utilization and subsequent loss. If this were the case, signs of chromium deficiency would be expected to be associated with increased physical training, which has not been observed. However, exercise also leads to increased insulin sensitivity of skeletal muscle glucose uptake and glycogen synthesis [20], increased ability to regulate blood glucose [21] and improved blood lipid profiles including increased HDL-cholesterol [22]. Improvements in these variables would likely mask any signs of marginal chromium deficiency.

Decreased basal chromium losses of trained subjects may reflect an adaptive mechanism leading to conservation of chromium stores. That postulate is supported by the work of Vallerand et al. [23] who reported that exercise-trained rats retained significantly higher chromium concentrations in the heart and kidneys compared with respective tissues of sedentary controls.

Adaptive mechanisms leading to conservation of chromium stores due to long-term exercise training may be important to trained athletes but would not affect the chromium losses of sedentary individuals who exercise strenuously but intermittently. These enhanced chromium losses, due to strenuous intermittent exercise, for individuals on calorie-restricted diets who exercise to lose weight would pose an even greater health problem. These individuals would display enhanced chromium losses, due to intermittent exercise accompanied by reduced chromium intake, leading to increased signs of marginal chromium deficiency.

Carbohydrate loading, a common practice among endurance athletes, often leads to increased muscle glycogen and increased physical performance [24–26]. Carbohydrate loading of 8 trained Navy divers led to increased work output and reduced chromium losses compared to those following the control diet period [27]. Urinary chromium losses are usually related to work output with more strenuous exercise causing greater chromium losses. Mild short-term exercise does not lead to detectable enhanced chromium losses.

Following carbohydrate loading, the relative stress associated with exercise was less than that following the controlled diet period based upon serum cortisol values [27]. Serum cortisol is related to the duration and intensity of exercise and therefore a measure of relative stress. For example, short-term running increased mean cortisol 27% while long-term exercise increased mean cortisol 43% [28].

The positive correlation between postexercise serum cortisol and urinary chromium losses is consistent with the observations that serum cortisol and urinary chromium losses are both increased with increasing amounts of stress. It appears that the decreased stresses associated with carbohydrate loading have a larger effect on indices of stress than those associated with increased duration of exercise. Both serum cortisol and urinary chromium losses appear to be meaningful indicators of overall physical stress.

Chromium Losses in Sweat

Loss of chromium in sweat has not been determined using acceptable collection and analytical techniques. This is an important area of research that needs to be explored especially in light of possible substantial losses of sweat associated with strenuous exercise. Effects of training and chromium supplementation on chromium sweat losses also need to be determined.

Exercise and Chromium – Animal Studies

Since stresses associated with exercise and high intakes of simple sugars enhance chromium losses, rats were raised on a low chromium high sugar diet to determine if exercise or training would exacerbate signs of chromium deficiency [29]. After 16 weeks of training, liver, gastrocnemius and biceps femoris glycogen concentrations were higher in the trained compared to the sedentary groups, independent of dietary chromium. There was a chromium-training interaction on glycogen synthase activities in liver and gastrocnemius muscle. Total protein concentration increased in the liver but decreased in the gastrocnemius due to dietary chromium.

Roginski and Mertz [30] reported that 18 h fasting liver glycogen levels were significantly lower in rats fed a low chromium diet compared to chromium-supplemented animals. Rats fed a low chromium diet also displayed decreased glycogen formation in liver and heart following insulin treatment compared with chromium-supplemented rats. Supplemental chromium also led to increased basal liver glycogen concentrations in turkey poults [31]. Chromium-supplemented turkeys also had greater glycogen synthase activities than birds fed the control diet.

These animal studies suggest that chromium increases glycogen concentrations. Since increased glycogen is associated with increased exercise performance, these studies suggest a possible role of sufficient dietary chromium in establishing optimal exercise performance.

There was also a significant chromium-training interaction on fasting serum cholesterol after 4, 8 and 12 weeks of training [32]. Sedentary animals consuming the low chromium diet had greater increases in cholesterol than the chromium-supplemented group relative to the trained groups.

Training did not consistently exacerbate signs of chromium deficiency associated with changes in glucose, insulin and triglycerides but did protect against increased serum cholesterol associated with consuming a low chromium diet.

Chromium and Exercise Performance

In the gastrocnemius muscle there was a significant effect of dietary chromium on muscle glycogen in response to acute exercise [33]. The drop in muscle glycogen was greater in the chromium-deficient rats than in the chromium-supplemented rats following 20 min of acute exercise. The low chromium rats retained approximately 24% of preexercise glycogen levels and the chromium-supplemented animals 41%. Since muscle glycogen is associated with exercise performance, the chromium-supplemented rats with greater glycogen reserves following acute exercise, should be able to exercise longer. Direct studies to measure exercise performance favored the plus chromium rats but results were inconclusive. Dietary chromium also has been shown to increase fasting liver glycogen suggesting a possible effect of dietary chromium on performance [30].

Chromium supplementation, in the form of chromium picolinate, led to significant increases in lean body mass of 10 subjects participating in weight training classes compared to control subjects [34]. Biceps and calf circumference also increased significantly in subjects receiving the chromium supplement compared to the placebo group. Actual increase in lean body mass of subjects on chromium was also greater than that of the placebo group. In a follow-up study involving college football players on a supervised weight-training schedule, increase in lean body mass of subjects given the chromium supplement exceeded those given the placebo by 44%. Decrease in total body fat was 3.5 times greater in men taking chromium than in placebo controls.

Insulin regulates the entry of glucose and amino acids into muscle cells and inhibits the action of enzymes that catabolize amino acids and proteins. Insulin also regulates lipid deposition in adipose cells [35]. By potentiating the action of insulin, the primary function of chromium, amino acid incorporation into muscle proteins also would be expected to increase leading to increased muscle mass. Efficient regulation of lipid deposition leading to decreased body fat also would be associated with the improved function of insulin. This increase in strength and decrease in percent body fat should lead to increased performance. Studies are in progress to document these observations.

Copper

Copper, like chromium, is involved in normal carbohydrate and lipid metabolism. Copper is also involved in immune function, collagen and elastin formation, amino acid metabolism, hematopoiesis and protection against cellular damage due to free radicals. Four copper-dependent enzyme systems play key roles in copper metabolism: (1) ceruloplasmin functions as a ferroxidase to improve iron utilization and prevent hematopoiesis; (2) the monoamine oxidase enzymes are involved in pigmentation and control of neurotransmitters and neuropeptides; (3) lysyl oxidase functions in maintaining structural integrity of connective tissue in lungs, bones and cardiovascular system, and (4) copper-dependent cytochrome oxidase plays a key role in the terminal steps of oxidative metabolism and superoxide dismutase is involved in the prevention of accumulation of superoxide radicals [36].

There has been a rapid expansion of research concerning the effects of marginal copper intake in animal studies that strongly suggests a key role of copper in cardiovascular diseases. However, these findings have not been appreciated by the medical community and marginal copper deficiency and human health remain largely separate entities.

Signs of Copper Deficiency, Dietary Intake and Requirement

Overt signs and symptoms of insufficient dietary copper include anemia, pancreatic atrophy, heart hypertrophy, glucose intolerance, elevated blood lipids and decreased ceruloplasmin, serum copper, hepatic copper and erythrocyte superoxide dismutase activity [36].

Dietary copper intake in most Western societies is below the suggested safe and adequate intake range of 1.5–3.0 mg/day [7]. Other components in the diet also have a strong influence on copper nutriture. Animal studies indicate that molybdenum, ascorbic acid, zinc, sulfur, iron, calcium, cadmium, silver, mercury, protein, carbohydrate and fiber effect copper metabolism [36]. Ascorbic acid [37] and zinc [36, 38] have been shown to effect the copper status of humans. The effects of dietary fiber and phytate, at amounts found in most human diets, are likely not significant problems in human nutrition.

Type of dietary carbohydrate has a profound effect in animal studies, e.g., dietary fructose strongly exacerbates signs of copper deficiency compared with glucose or starch [39]. Studies are in progress to document this work in humans. One study to document a carbohydrate effect on copper metabolism in humans had to be terminated prematurely due to cardiac abnormalities in 4 of 23 subjects due apparently to the low copper diets employed, 0.35 mg/1,000 cal [40]. Copper content of those diets was similar to that consumed by a significant portion of the normal population. Cardiac abnormalities included one coronary infarct, two incidences of severe tachycardia and one incidence of extrasystolic beats. That study may indicate that increased stress whether due to diet, physical exercise or unknown variables may lead to possible overt signs of copper deficiency in humans.

Foods high in copper include liver, cereal products, prunes, raisins and nuts. Poor dietary sources include dairy products and refined simple sugars such as sucrose, fructose and lactose [36]. In humans, copper toxicity is characterized by metallic taste, epigastric pain, nausea, vomiting, diarrhea and in severe cases vascular collapse and death. Due to the stronger interactions of copper with other components of the diet, specific toxic levels cannot be defined [36].

Exercise and Blood Copper

In plasma or serum most of the copper (90–93%) is bound to ceruloplasmin and the remaining copper is bound less firmly to albumin and a smaller percentage to amino acids [41]. In normal blood approximately half the copper is present in erythrocytes where 40% of the copper is bound loosely to amino acids and 60% bound to superoxide dismutase (SOD). SOD is used as a measure of copper status in humans [42] and is a better measure of copper status than total blood copper [43].

SOD is a metalloenzyme with copper involved directly in enzyme activity and zinc in a structural role [44]. This metalloenzyme helps destroy free radicals, including the superoxide anion, that are produced during exercise in muscles and liver. SOD activity in skeletal muscle increases with increased peak oxygen uptake [45] and is also increased in male and female swimmers during the training season [46] without any changes in serum copper [47]. Since SOD activity is related to copper status, it is essential that exercising individuals maintain proper copper nutrition.

Male university athletes displayed higher plasma copper levels than subjects who were not members of university athletic teams [48]. Increased resting plasma copper levels were reported for 31 male and female elite runners, 10–17 years, compared with 21 age-matched controls [49]. In a separate study, mean copper levels were similar for runners and nonrunners and there was no relationship between training mileage and serum copper [50]. Singh et al. [51] reported higher mean plasma copper concentrations of female runners compared with nonrunners (18.8 ± 0.5 vs. 16.1 ± 0.6 µmol/l) but erythrocyte copper was significantly lower (1.06 ± 0.02 vs. 1.26 ± 0.03 µg/g). Anderson et al. [52] reported that serum copper concentrations of trained subjects tended to be higher than those for untrained subjects but differences were not significant. Differences between runners and nonrunners suggest that chronic exercise may induce a redistribution of tissue copper similar to that observed for chromium and zinc.

Blood copper, in response to acute exercise, has been reported to increase [53–55], decrease [56] or remain constant [16]. Increases in serum copper were 35% for trained athletes compared with 15% increase in untrained athletes following 90 min of intense exercise on an ergometer [54]. Type of exercise, e.g., cycle ergometer, swimming, or running as well as degree of training are likely to affect the acute changes in blood copper associated with exercise.

Copper Losses and Exercise

Most absorbed copper is excreted into the bile and ultimately excreted in the feces. Only a small portion (1–2%) of absorbed copper is excreted in the urine with larger amounts lost in gastric secretions and sweat. Copper absorption in humans varies from 25 to 70% with most of the absorption occurring in the small intestine but some absorption has been demonstrated for the stomach.

Exercise leads to substantial increases in urinary losses of chromium and zinc but urinary copper losses are largely unaffected by exercise [16]. Basal urinary excretion of trained individuals, 9.5 ± 0.06 µg/day, was lower than that of untrained subjects, 12.2 ± 0.09 µg/day [17]. Changes in fecal copper due to exercise are difficult to assess since feces contain not only nonabsorbed copper but also absorbed copper that was excreted into the feces via the bile as well as copper sloughed off the gastrointestinal tract [36].

Copper Losses in Sweat

Sweat loss may be 2–4 liters/h when competing in endurance events [57]. Copper concentrations of sweat vary from 58 to 500 µg/l. Large variations in copper concentration in sweat may be due to analytical problems, contamination, copper intake and status of subjects, area of body sweat was collected, degree of training of subjects, and methods and conditions used to induce sweating [2].

Normal sweat losses of copper approximate 25–30% of the dietary requirement for this metal [58]. Three male subjects fed a constant copper diet containing 3.5 mg copper/day in a controlled environment (37.8 °C, 50% humidity) were in negative copper balance during 10 days of observation. The average sweat loss of copper was 1.6 mg/day [59]. Concentrations of copper from abdomen tended to be the highest (14.0 ± 9.9 µmol/l) followed by that from the chest (11.5 ± 11.6 µmol/l), back (8.8 ± 6.8 µmol/l) and arm (8.2 ± 7.5 µmol/l) but due to the large variations among subjects, differences were not significant [60]. Low molecular mass complexes of copper detectable by their ability to bind o-phenanthroline were found at concentrations much greater in arm than trunk sweat [61]. Copper complexes in sweat samples significantly stimulated lipid peroxidation. It is possible that these potentially damaging copper complexes may be excreted in the sweat as a protective mechanism thereby diminishing the extent of free radical damage in vivo.

Exercise and Copper – Animal Studies

Ceruloplasmin, a copper-binding protein, functions in the formation of the Fe(III)-transferrin complex and is therefore important in iron transport and availability. Exercise (swimming) in rats, like in humans, leads to increased ceruloplasmin levels [62]. Serum ceruloplasmin activity was also significantly increased in trained (running) male rats compared with untrained controls. Exhaustive exercise also led to significantly elevated ceruloplasmin in both trained and untrained rats [63].

The increase in ceruloplasmin in response to exercise may be a nonspecific response to the stress of exercise since various stresses increase ceruloplasmin levels. Exercise-induced increases in ceruloplasmin also may reflect increased requirement for copper and iron transport required during exercise. Exercise (swimming) also tended to increase copper in spleen, liver and heart of male rats [62]. There were no effects of exercise in female rats. Whether the different responses of males and females are due to sexual differences or a relative lack of swimming in the female rats that floated easily is not known.

There are a limited number of animal studies reporting the effects of exercise on copper status and variables in experimental animals. Due to the important role of copper in exercise metabolism, this is an important area of study that requires additional research.

Copper and Exercise Performance

While there is no direct evidence linking dietary copper directly with exercise performance, performance is linked to insufficient dietary iron [3]. The formation of Fe(III)-transferrin complex that is involved in iron transport and availability is a copper-dependent process (see section on Exercise and Copper – Animal Studies). Therefore, some of the effects of iron on exercise performance may be indirect copper-dependent effects on performance.

Free radical generation is increased during strenuous exercise [64] and copper plays a key role in removal of free radicals, e.g., SOD. The effects of free radicals may not be detectable for years, e.g., in cardiovascular diseases. Therefore, the effects of proper copper nutrition may be in the alleviation of signs and symptoms associated with cardiovascular diseases that may take years to develop. Much more work is required to substantiate the role of copper in these diseases associated with free radical damage.

Zinc

Zinc, a component of more than 100 enzymes, is found in the nuclear, mitochondrial and supernatant fractions of all cell types. It is a multifunctional nutrient involved in cell replication and differentiation, glucose and lipid metabolism, hormone function, growth, taste, immune function, sexual maturation, reproduction, wound healing, and vision. Zinc is involved also in energy production in muscles as well as removal of lactic acid making it a key component in the exercising muscle. Fifty to 60% of total body zinc (2 g) is present in skeletal muscle of humans [65].

Signs of Zinc Deficiency, Dietary Intake and Requirement

Signs of marginal zinc deficiency observed in humans include slow wound healing, anorexia, oligospermia, loss of taste and smell, decreased growth and decreased immune function. Excess zinc intake leads to anemia, decreased HDL-cholesterol, diarrhea, nausea, vomiting, headaches, lethargy and copper deficiency [65].

The recommended daily allowances for adult males is 15 mg and 12 mg for females [7]. However, dietary zinc intake is usually below the RDA. Zinc is absorbed at a rate of 10–40% primarily in the duodenum and jejunum with lesser amounts in the ileum. Good sources of dietary zinc include oysters, beef, liver, dark meat from chicken and turkey. Grains, cereals, nuts and legumes also contain significant amounts of zinc but often of lower bioavailability [65].

Exercise Effects on Blood Zinc

Serum zinc is often used to assess zinc status but is not a sensitive specific indicator of zinc status. Ratio of serum zinc and the sum of the serum concentrations of zinc carriers has been used as a better indicator [67]. Serum zinc has been reported by most researchers to decline with endurance training in runners [50, 66, 67] and wrestlers [68]. Intensive 34-day training maneuvers also caused a decline in serum zinc levels of 30 US Army soldiers [69]. Declines in serum zinc may be related to increased sweat losses associated with strenuous exercise. Decreases in albumin and alpha-2-macroglobulin, two zinc-binding proteins associated with exercise, also may explain decreased blood zinc values [67]. Relocalization of zinc in tissues would also explain decreased blood zinc in trained individuals.

Fasting plasma zinc of highly trained women runners was similar to that of untrained women [70]. No differences in plasma zinc levels were found between 44 male university athletes and 20 male students not in athletic teams [48]. However, both groups were in good physical condition based upon their V_{O_2max} values. Plasma zinc of male and female swimmers also did not change during the training season [47]. Serum zinc of trained and untrained subjects based on V_{O_2max} values was also similar [52]. Therefore, exercise as well as other factors including dietary zinc intake including supplementation, intake of other nutrients that effect zinc metabolism and duration and type of exercise need to be considered to evaluate zinc status of trained athletes.

Acute exercise usually results in increased serum or plasma zinc immediately following exercise depending upon type of exercise and physical status of subjects [48, 52, 56, 71]. Immediately following acute exercise on a treadmill at 90% of V_{O_2max} to exhaustion (30 s exercise and 30 s rest periods), serum zinc was significantly elevated in both trained and untrained subjects [52]. Immediately following a 10-km run, serum zinc was insignificantly higher, 85 ± 4 vs. 81 ± 4 µg/dl, before race and declined significantly to 75 ± 4 µg/dl 2 h following the run [16]. With brief exercise (10-min running stairs) there was a rapid increase in plasma zinc followed by a decline. After 1 h, both plasma zinc and albumin had fallen but albumin returned to preexercise values while plasma zinc fell below preexercise values. Prolonged exercise (10 miles at 8 ± 1 min/mile) resulted in decreased plasma zinc that was still significantly lower than preexercise values 6 h postexercise. Albumin levels remained at preexercise levels [72]. Serum zinc concentrations in men participating in a 70-km cross-country ski race lasting 5 h were 19% higher immediately following the race compared to prerace values and returned to prerace levels within 1 day [71].

The rapid postexercise drop in serum zinc levels is due to increased urinary losses coupled with a redistribution of zinc from blood to tissues, primarily the liver, in response to increased stress responses. Leukocyte endogenous mediator, a hormone-like substance released by phagocytes in response to stress, leads to a redistribution of zinc from blood to liver [2]. A stable factor seems to be present in blood following exercise, since human plasma collected within 3 h of exercise depressed plasma zinc when injected intraperitoneally into rats [73]. Stress, including exercise, induces the liver to produce metallothionein, a zinc-binding protein [74], which may also alter tissue distribution of zinc.

Zinc Losses and Exercise

Most of the absorbed zinc is lost in pancreatic and intestinal secretions that are subsequently eliminated in the faeces. Normal urinary zinc losses are 300–600 µg/day; various forms of stress, including exercise, increase urinary zinc losses [65].

Urinary zinc losses usually increase in response to acute strenuous exercise and increases are dependent upon duration and intensity of exercise. Urinary zinc losses of male runners increased almost 50% from 489 to 711 µg/day on the day of a 10-km run compared to a nonrun day. Urinary losses of trained runners on a controlled diet increased insignificantly following exercise at 90% of V_{O_2max} to exhaustion with 30s exercise and 30 s rest periods [12]. Urinary zinc excretion increased transiently with minimal effects on daily losses following a 10-mile run at 8 ± 1 min/mile [72]. Highly trained women runners had significantly higher urinary zinc losses than untrained women [70]. Urinary losses of male trained runners were also greater than untrained controls but differences were not significant [12]. The greater urinary zinc excretion may reflect increased muscle breakdown and subsequent muscle turnover. Urinary losses of zinc by soldiers on maneuvers also increased but there was also a decrease in total body weight [69]. Muscle protein catabolism that occurs during weight loss [75] may contribute to the increased urinary losses in addition to the increased urinary zinc losses specific for exercise.

Zinc Losses in Sweat

Heavy sweating in hot humid environment leads to significant losses of total body zinc and may contribute to marginal zinc deficiency [58, 59]. However, there is an adaptation to sweat zinc losses leading to decreased zinc in sweat associated with acclimatization [59] or marginal zinc deficiency [76]. Normal surface losses of zinc were estimated at 4% in a group of nontraining men in a metabolic unit [58]. Concentration of sweat collected from chest, arm, back and abdomen varied from 6.4 to 12.7 µmol/l (416–825 µg/l) [60]. Assuming a sweat loss of 2–4 liters/h when competing in endurance events [57] and an average zinc concentration in sweat of 620 µg/l, total sweat losses would be 1,240–2,480 µg/h. Assuming an intake of 15 mg and an absorption of 10% (see section on Signs of Zinc Deficiency, Dietary Intake and Requirement), the losses of zinc in sweat in 1 h of strenuous exercise would exceed the 1.5 mg of zinc absorbed per day (10% of 15 mg daily intake). Even at a zinc absorption of 40%, the 2.5 mg lost in sweat would be over 40% of the total absorbed zinc of 6 mg (40% of 15 mg dietary intake). As is obvious from these calculations, zinc losses in sweat are substantial and are of particular importance to those who exercise strenuously and lose copious amounts of sweat.

Sweating is usually not associated with swimming but dehydration and weight loss do occur due to sweating. Swimmers may represent a group of athletes at increased risk of impaired trace element nutriture due to accelerated surface losses of micronutrients associated with prolonged training in water [47].

Exercise and Zinc – Animal Studies

An effect of exercise on zinc metabolism was first reported in dogs that displayed a marked increase in serum zinc concentrations following short bouts of intense exercise [77]. Zinc is also involved in muscle growth and function. Rats fed zinc-deficient diets have decreased muscle growth and DNA concentrations [78]. Richardson and Drake [79] reported that gastrocnemius muscle from rats fed supplemental zinc took longer to fatigue than that from control animals. Increased twitch tension in the living frog sartorius muscle was also observed following perfusion with $0.1\ M$ zinc solution [80]. These data suggest an effect of zinc on overall performance in animals. However, McDonald and Keen [3] could not find any difference in endurance capacity of rats fed either a zinc-deficient or adequate diet despite significant differences in serum zinc concentrations between the two groups (122 vs. 65 µg/dl).

Zinc and Exercise Performance

Due to the ubiquitous nature of zinc in exercise-dependent reactions, insufficient dietary zinc is a likely potential cause for suboptimal exercise performance in humans as well as experimental animals (see above). Marginal zinc deficiency effects zinc metalloenzymes in muscles, protein catabolism is increased leading to muscle fiber loss especially in muscle groups with higher zinc concentrations [72]. Exercise leads to changes in both intra- and extracellular zinc concentrations leading to altered zinc metabolism. Dynamic (isokinetic) strength and isometric endurance increased in 16 female volunteers [81] following 14 days of supplemental zinc (135 mg/day). However, additional studies are needed to document an effect of supplemental zinc on performance in humans.

Summary and Conclusion

In summary, dietary intake of the trace elements, chromium, copper and zinc is suboptimal based on the suggested safe and adequate intakes or recommended dietary allowance. These trace elements play key roles in energy production and utilization and are therefore of utmost importance in exercise performance. Strenuous exercise leads to altered trace metal distribution and increased losses in urine, faeces and sweat. There are no documented reports demonstrating that the overall trace metal status of athletes who exercise strenuously is significantly different from that of sedentary individuals. However, several studies suggest that trace metal status of athletes may be compromised. While numerous studies suggest possible benefits of trace elements on exercise performance, there are few studies documenting significant increases in athletic performance. Most studies attempting to document an effect of nutrient supplementation on exercise performance involve unbalanced incomplete nutrient supplements and may lead to erroneous conclusions. Supplementation of individual nutrients also may not yield meaningful answers since diets that are low in one trace element are likely low in others as well. A more meaningful study would be to supplement athletes with a complete balanced nutrient supplement in a double-blind well-controlled study employing a suitable number of subjects to yield statistically significant results. Top athletes are often performing near their peak levels and only small improvements would be expected. Athletes and coaches are also urged not to take or suggest excessive amounts of any nutrients as this will usually lead to detrimental effects. The National Academy of Sciences [7] assembles outstanding nutrition researchers to establish suitable intakes of nutrients and it is advisable to stay within their established limits. Trace elements should not be considered the panacea of optimal health or performance but should be considered as key elements to overall health both in exercising and sedentary individuals.

References

1 Campbell WW, Anderson RA: Effects of aerobic exercise and training on the trace minerals, chromium, zinc and copper. Sports Med 1987;4:9–18.
2 Anderson RA, Guttman HN: Trace minerals and exercise; in Terjung R, Horton ES (eds): Exercise, Nutrition and Energy Metabolism. New York, Macmillan, 1988, pp 180–195.
3 McDonald R, Keen CL: Iron, zinc and magnesium nutrition and athletic performance. Sports Med 1988;5:171–184.
4 Tuman RW, Bilbo JT, Doisy RJ: Comparison and effects of natural and synthetic glucose tolerance factor in normal and genetically diabetic mice. Diabetes 1978;27: 49–56.
5 Anderson RA: Essentiality of chromium in humans. Sci Total Environ 1989;86: 75–81.
6 Polansky MM, Anderson RA: Metal-free housing units for trace element studies involving rats. Lab Anim Sci 1979;29:357–359.
7 National Academy of Sciences: Recommended Dietary Allowances, ed 10. Washington, National Academy Press, 1989.
8 Anderson RA, Kozlovsky AS: Chromium intake, absorption and excretion of subjects consuming self-selected diets. Am J Clin Nutr 1985;41:1177–1183.
9 Bunker W, Lawson MM, Delves HT, et al: The uptake and excretion of chromium by the elderly. Am J Clin Nutr 1984;39:799–802.
10 Koivistoinen P: Mineral element composition of Finnish foods: N, K, Ca, Mg, P, S, Fe, Cu, Mn, Zn, Mo, Co, Ni, Ca, F, Se, Si, Rb, Al, B, Br, Hg, As, Cd, Pb and ash. Acta Agric Scand 1980;suppl 22.
11 Gibson RS, Scythes CA: Chromium, selenium, and other trace element intakes of a selected sample of Canadian premenopausal women. Biol Trace Elem Res 1984;6: 105–116.
12 Anderson RA, Polansky MM, Bryden NA, et al: Chromium requirement is related to degree of glucose intolerance; in Momčilović B (ed): Seventh International Symposium on Trace Elements in Man and Animals, Dubrovnik, Yugoslavia 1991, vol 7, in press.
13 Anderson RA, Bryden NA: Concentration, insulin potentiation and absorption of chromium in beer. J Agric Food Chem 1983;31:308–311.
14 Kozlovsky AS, Moser PB, Reiser S, et al: Effects of diets high in simple sugars on urinary chromium losses. Metabolism 1986;35:515–518.
15 Anderson RA: Chromium; in Mertz W (ed): Trace Elements in Human and Animal Nutrition, ed 5. Orlando, Academic Press, 1987, vol 1, pp 225–244.

16   Anderson RA, Polansky MM, Bryden NA: Strenuous running: Acute effects on chromium, copper, zinc and selected clinical variables in urine and serum of male runners. Biol Trace Elem Res 1984;6:327–336.
17   Anderson RA, Bryden NA, Polansky MM, et al: Exercise effects on chromium excretion of trained and untrained men consuming a constant diet. J Appl Physiol 1988; 64:249–252.
18   Anderson RA, Polansky MM, Bryden NA, et al: Effect of exercise (running) on serum glucose, insulin, glucagon and chromium excretion. Diabetes 1982;31:212–216.
19   American Heart Association: Exercise testing and training of apparently healthy individuals; in Handbook for Physicians, 1972.
20   Richter EA, Ploreg T, Galbo H: Increased muscle glucose uptake after exercise. Diabetes 1985;34:1041–1048.
21   Garetto LP, Richter EA, Ruderman NB: Enhanced muscle glucose metabolism after exercise in rat: the two phases. Am J Physiol 1984;246:E471–E475.
22   Dufaux B, Assmann G, Hollman W: Plasma lipoproteins and physical activity: a review. Int J Sports Med 1982;3:123–136.
23   Vallerand AL, Cuerrier JP, Shapcott D, et al: Influence of exercise training on tissue chromium concentrations in the rat. Am J Clin Nutr 1984;39:402–409.
24   Christensen EH, Hansen O: Respiratorischer Quotient und O_2 Aufnahme. Skand Arch Physiol 1939;81:180–189.
25   Bergstrom J, Hultman E: A study of the glycogen metabolism during exercise in man. Scand J Clin Lab Invest 1967;19:218–228.
26   Hultman E: Studies on muscle metabolism of glycogen and active phosphate in man with special reference to exercise and diet. Scand J Clin Lab Invest 1967;19:94–99.
27   Anderson RA, Bryden NA, Polansky MM, et al: Carbohydrate loading and exercise effects on chromium and zinc losses and serum cortisol and insulin. FASEB J 1990; 4:2960.
28   Kuoppasalmi K, Naveri H, Harkonim M, Aldercreutz H: Plasma cortisol, aldrostinedione, testosterone and luteinizing hormone in running exercise of different intensities. Scand J Clin Lab Invest 1980;40:403–409.
29   Campbell WW, Polansky MM, Bryden NA, et al: Exercise training and dietary chromium effects on glycogen, glycogen synthase, phosphorylase and total protein in rats. J Nutr 1989;119:653–660.
30   Roginski EE, Mertz W: Effects of chromium (III) supplementation on glucose and amino acid metabolism in rats fed a low protein diet. J Nutr 1969;97:525–530.
31   Rosebrough RW, Steele NC: Effect of supplemental dietary chromium or nicotinic acid on carbohydrate metabolism during basal, starvation, and refeeding in poults. Poult Sci 1981;60:407–417.
32   Campbell WW, Polansky MM, Bryden NA, et al: Dietary chromium and exercise training effects on glucose, cholesterol and related variables. J Trace Elem Exp Med 1990;3:291–305.
33   Campbell WW: Effects of exercise training and dietary chromium on glucose/glycogen metabolism, chromium retention and selected clinical variables; masters thesis, University of Maryland, College Park, 1987.
34   Evans GW: The effect of chromium picolinate on insulin controlled parameters in humans. Int J Biosoc Med Res 1989;11:163–180.

35 Felig P: Amino acid metabolism in man. Annu Rev Biochem 1975;44:933–955.
36 Davis GK, Mertz W: Copper; in Mertz W (ed): Trace Elements in Human and Animal Nutrition. Orlando, Academic Press, 1987, vol 1, pp 301–364.
37 Finley EB, Cerklewski FL: Influence of ascorbic acid supplementation on copper status in young adult men. Am J Clin Nutr 1983;37:553–556.
38 Goodwin JS, Hunt WC, Hooper P, et al: Relationship between zinc intake, physical activity and blood levels of high-density lipoprotein cholesterol in a healthy elderly population. Metabolism 1985;34:519–523.
39 Fields M, Ferretti RJ, Smith JC, et al: The interaction of type of dietary carbohydrates with copper deficiency. Am J Clin Nutr 1984;39:289–294.
40 Reiser S, Smith JC Jr, Mertz W, et al: Indices of copper status in humans consuming typical American diet containing either fructose or starch. Am J Clin Nutr 1985;42:242–251.
41 Neuman PZ, Sass-Kortsak AI: The state of copper in human serum: Evidence for an amino acid-bound fraction. Clin Invest 1967;46:646–658.
42 Bennett FI, Golden MHN, Golden BE, et al: Red cell superoxide dismutase as a measure of copper status in man; in Mills CF, Bremner I, Chesters JK (eds): Commonwealth Agric Bureaux, Farnham Royal, UK, 1985, vol 5, pp 578–581.
43 Masters HG, Smith GM, Casey RH: The relationship between the activity of superoxide dismutase and the concentration of Cu in the levels of molybdenum; in Mills CF, Bremner I, Chesters JK (eds): Commonwealth Agric Bureau, Farnham Royal, UK, 1985, vol 5, pp 575–577.
44 McCord JM, Fridovich I: Superoxide dismutase, an enzymatic function for erythrocuprein (hemocurperin). J Biol Chem 1969;244:6049–6055.
45 Jenkins RR, Friedland R, Howard H: The relationship of oxygen uptake to superoxide dismutase and catalase activity in human skeletal muscle. Int J Sports Med 1984;5:11–14.
46 Lukaski HC, Hoverson BS, Milne DB, et al: Copper, zinc, and iron status of female swimmers. Nutr Res 1989;9:493–502.
47 Lukaski HC, Hoverson BS, Gallagher SK, et al: Physical training and copper, iron and zinc status of swimmers. Am J Clin Nutr 1990;51:1093–1099.
48 Lukaski HC, Bolonchuk LM, Klevay DB, et al: Maximal oxygen consumption as related to magnesium, copper and zinc nutriture. Am J Clin Nutr 1983;37:407–415.
49 Conn CA, Schemmel RA, Ku P, et al: Relationship of maximal oxygen consumption to plasma and erythrocyte magnesium and to plasma copper levels in elite young runners and controls. Fed Proc 1986;45:972.
50 Dressendorfer RH, Sockolov R: Hypozincemia in runners. Phys Sports Med 1980;8:97–100.
51 Singh A, Deuster PA, Moser PB: Zinc and copper status of women by physical activity and menstrual status. J Sport Med Phys Fitness 1990;30:29–36.
52 Anderson RA, Bryden NA, Polansky MM, et al: Exercise effects on urinary losses and serum concentrations of chromium, copper, iron and zinc of trained and untrained runners. FASEB J 1989;3:1294.
53 Haralambie G: Changes in electrolytes and trace elements during exercise; in Howard H, Poortmans JR (eds): Metabolic Adaptation to Prolonged Physical Exercise. Basel, Birkhäuser, 1975, pp 340–351.

54 Olha AE, Klissouras V, Sullivan JD: Effect of exercise on concentration of elements in serum. J Sports Med 1982;22:414–425.
55 Ohno H, Yamashita K, Doi R, et al: Exercise-induced changes in blood zinc and related proteins in humans. J Appl Physiol 1985;58:1453–1458.
56 Rusin VY, Nasalodin VV, Varobev VA: Iron, copper, manganese and zinc metabolism in athletes under high physical pressure. Vapr Pitan 1980;4:15–19.
57 American Dietetic Association: Nutrition and physical fitness. J Am Diet Assoc 1980;76:437–443.
58 Jacob RA, Sandstead HH, Munoz JM, et al: Whole body surface loss of trace metals in normal mates. Am J Clin Nutr 1981;34:1379–1383.
59 Consolazio CF, Nelson RA, Matoush RC, et al: The trace element mineral losses in sweat. Denver, US Army Medical Research and Nutrition Laboratory 1964, Rep No 284.
60 Aruoma OI, Reilly T, MacLaren D, et al: Iron, copper and zinc concentrations in human sweat and plasma; effect of exercise. Clin Chim Acta 1988;177:81–88.
61 Gutteridge JMC, Rowley DA, Halliwell B, et al: Copper and iron complexes catalytic for oxygen radicals in sweat from human athletes. Clin Chim Acta 1985;145:267–273.
62 Ruckman KS, Sherman AR: Effects of exercise on iron and copper metabolism in rats. J Nutr 1981;111:1593–1601.
63 Dowdy RP, Dohm GL: Effect of training and exercise on serum ceruloplasmin in rats. Proc Soc Exp Biol Med 1972;139:489–491.
64 Demopoulos HB, Santomier JP, Seligman ML, et al: Free radical pathology: rationale and toxicology and antioxidants and other supplements in sports medicine and exercise science; in Katch FI (ed): Sport, Health and Nutrition: The 1984 Olympic Scientific Congress Proceedings. Champaign, Human Kinetics Publishers, 1984, vol 2, pp 139–189.
65 Hambidge KM, Casey CE, Krebs NF: Zinc; in Mertz W (ed): Trace Elements in Human and Animal Nutrition, ed 5. Orlando, Academic Press, 1987, pp 1–138.
66 Haralambie G: Serum zinc in athletes in training. Int J Sports Med 1981;2:135–138.
67 Couzy F, Lafargue P, Guezennec CY: Zinc metabolism in the athlete: influence of training, nutrition and other factors. Int J Sport Med 1990;11:263–266.
68 Oberleas D, White RC, Hurley LS, et al: Evidence for attractions of zinc and magnesium in conditioned athletes. Fed Proc 1972;31:68.
69 Miyamura JB, McNutt SW, Lichton IJ, et al: Altered zinc status of soldiers under field conditions. Res 1987;87:595–597.
70 Deuster PA, Day BA, Singh A, et al: Zinc status of highly trained women runners and untrained women. Am J Clin Nutr 1989;49:1295–1301.
71 Hetland O, Brubak EA, Refsum HE, et al: Serum and erythrocyte zinc concentrations after prolonged exercise; in Howard H, Poortmans J (eds): Metabolic Adaptation to Prolonged Physical Exercise. Basel, Birkhäuser, 1975, pp 367–370.
72 Van Rij AM, Hall MT, Dohm GL, et al: Changes in zinc metabolism following exercise in humans. Biol Trace Elem Res 1986;10:99–105.
73 Pekarek RS, Wannemacker RW, Beisel WR: The effect of leucocytic endogenous mediator in the tissue distribution of zinc and iron. Proc Soc Exp Biol Med 1972;140:685–688.

74 Oh SH, Deagen JT, Whanger PD, et al: Biological function of metallothionein. V. Its induction in rats by various stresses. Am J Physiol 1978;234:E282–E285.
75 Spencer H, Kramer L, Perakis E, et al: Plasma levels of zinc during starvation. Fed Proc 1982;41:347.
76 Prasad AS, Schubert AR, Sandstead HH, et al: Zinc, iron and nitrogen content of sweat in normal and deficient subjects. J Lab Clin Med 1963;62:84–89.
77 Lichti E, Turner M, Deweese M, et al: Zinc concentration in venous plasma before and after exercise in dogs. Mo Med 1970;303–304.
78 Park JH, Grandjean CJ, Antonson DL, et al: Effects of isolated zinc deficiency on composition of skeletal muscle, liver and bone during growth in rats. J Nutr 1986;116:610–617.
79 Richardson JH, Drake PD: The effects of zinc on fatigue of striated muscle. J Sports Med 1979;19:133–134.
80 Isaacson A, Sandow A: Effects of zinc on responses of skeletal muscle. J Gen Physiol 1963;46:655–677.
81 Krotkiewski M, Gudmundsson M, Backstrom P, et al: Zinc and muscle strength. Acta Physiol Scand 1982;116:309–311.

Richard A. Anderson, PhD, Vitamin and Mineral Nutrition Laboratory,
Beltsville Human Nutrition Research Center, US Department of Agriculture, ARS,
Beltsville, MD 20705 (USA)

Exercise and Free Radicals: Effects of Antioxidant Vitamins

Adrianne Bendich

Department of Clinical Nutrition, Hoffmann-La Roche Inc., Nutley, N.J., USA

Introduction

Skeletal muscle and brain have the highest rates of oxidative metabolism of all human tissues [24]. Exercise, by its very nature, increases oxygen consumption and thus indirectly enhances the potential for the formation of oxygen-containing free radicals. The highly energized radicals can damage muscle tissue and have adverse systemic effects. Antioxidants are capable of lowering the free radical burden associated with strenuous exercise. This chapter briefly reviews the major sites and effects of free radical generation in muscle; methodologies available for measurement of the products of free radical reactions; and examines the protective effects of antioxidant vitamins, especially vitamin E. The role of glutathione, the antioxidant enzymes and other antioxidants of particular importance to skeletal muscle are also examined. The inconsistent findings of effects of the antioxidant vitamins on muscle strength and/or performance have been reviewed by Williams [50, 51] and Gerster [14] and are not discussed in this review.

Free Radicals

Endogenous Sources
Strenuous physical exercise is associated with increases in the size and number of mitochondria in the exercised muscle. Mitochondrial enzyme activities associated with the respiratory chain (occurring in the inner membrane) are also enhanced [reviewed in 18]. Over 98% of oxygen uti-

lized by cells is efficiently reduced by the mitochondrial electron transport system, resulting in the formation of ATP and water. The remaining oxygen is available for one or two electron reduction, resulting in the formation of the superoxide free radical and hydrogen peroxide. These two products can react to also form water and the highly reactive hydroxyl radical. Cytochrome oxidase activity, the final enzyme involved in electron transport, is increased over 30% in exercise-trained versus sedentary rats [46]. Electron transport systems are thus continuous sources of intracellular, reactive oxygenated free radicals, defined as compound(s) or element(s) with one or more unpaired electrons [31].

The mitochondrial matrix is the site of fatty acid oxidation, oxidative decarboxylation of pyruvate and the reactions involved in the citric acid cycle. Even though these and the respiratory chain reactions occur at different sites in the mitochondrion, free radical damage to the inner membranes can increase the activities of cytochrome c reductases; enzymes associated with oxidative pathways and lower the production of ATP [16]. For example, muscle homogenates from exercised rats show increased levels of free radical production as determined by electroparamagnetic resonance signals and concomitant reduced mitochondrial respiratory control compared to skeletal muscle homogenates from sedentary rats [10]. Electrical stimulation of contractions in whole muscle from rats was associated with a 70% increase in the free radical signal compared to that seen from unstimulated muscle [21] as well as a significant increase in muscle enzyme release.

Muscle fiber sites of free radical generation, in addition to mitochondria, include lysosomes, peroxisomes, nuclear and sarcoplasmic reticulae and the sarcolemma (fig. 1). Free radicals and their products are also present within the sarcoplasm. Molecular species include, but are not limited to hydroxyl, peroxy, superoxide and alkoxy radicals and reactive molecules such as hydrogen peroxide and singlet oxygen, which are not free radicals but are reactive and capable of causing cell damage [31].

Major targets of free radical damage are the unsaturated bonds in cellular membrane lipids [45] (fig. 1). Muscle fibers contain high concentrations of arachidonic acid in their phospholipids as well as specific cytochrome-linked fatty acid desaturase systems [39]. The prostaglandin E_2 (PGE_2) product of arachidonic acid oxidation is an important regulator of muscle protein turnover [20]. Products of lipoxygenase activation are chemotactic and stimulate cells associated with inflammatory responses to move to the damaged muscle. Endurance training has been shown to

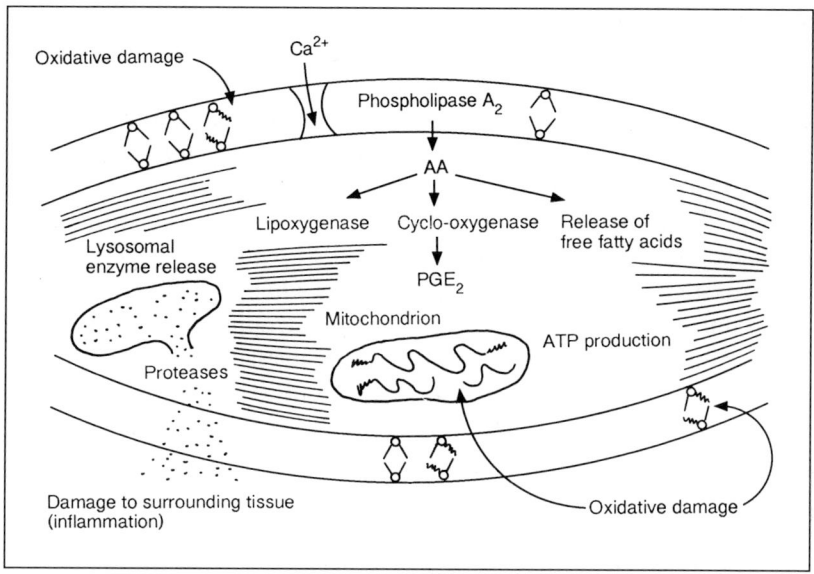

Fig. 1. Potential sites of oxidative damage to skeletal muscle including increased influx of calcium ions, which can activate PLA_2 and/or proteases; free radical damage to muscle membranes including the sarcolemma, mitochondria and lysosomes.

increase the capacity of skeletal muscle to oxidize long chain fatty acids, such as arachidonic acid [18].

Lipid peroxidation can result in a loss of membrane fluidity, receptor alignment, ion channel permeability and potentially cellular lysis. Membranes become leaky, which is reflected in an increased serum concentration of muscle-specific enzymes (such as pyruvate kinase). Leakage of lysosomal enzymes, such as proteases, can significantly disrupt skeletal muscle contractility. Free radical damage to the sulfur-containing enzymes found in high concentrations in skeletal muscle may result in inactivation, crosslinking and/or denaturation. In particular, skeletal muscle contains a high concentration of the amino acid phenylalanine (approximately 4% of muscle tissue) which is highly sensitive to attack by hydroxyl radicals [42].

Calcium channels can be altered, resulting in calcium influx, activation of phospholipase A_2 (PLA_2) and consequent initiation of the arachidonic acid cascade (fig. 1). Nucleic acids can be attacked and DNA damaged. Intracellular free radicals are also generated from the auto-oxidation and consequent inactivation of small molecules such as thiols and catecholamines; the activity of certain oxidases such as xanthine oxidase which is present in skeletal muscle [25], cyclo-oxygenases, lipoxygenases, dehydrogenases, peroxidases and several other metabolic pathways [20, 31].

In addition to the continuous, low level production of reactive oxygen intermediates during normal cellular metabolism, higher concentrations of free radicals are utilized as part of the first line of defense against pathogens by white blood cells (leukoytes). Neutrophils, the most abundant white blood cells, have the capacity to take up molecular oxygen and generate reactive oxygen-containing molecules when stimulated. This is often called the oxidative or respiratory burst. Neutrophils and other phagocytic leukocytes can also generate highly toxic halogenated molecules [45]. Neutrophils are found associated with injured and/or inflamed tissues. Skeletal muscle which, during strenuous exercise, undergoes severe physical stresses including shearing forces, often becomes inflamed [20]. As mentioned above, certain products of muscle membrane lipid peroxidation are chemoattractants for neutrophils.

Exogenous Sources

Exogenous sources of free radicals include radiation (from x-rays, radon, ultraviolet light), tobacco smoke, certain air pollutants such as ozone and nitrogen dioxide and the products of automobile exhaust systems, organic solvents, anesthetics, hyperoxic environments and pesticides. Some of these compounds as well as certain medications are metabolized to free radical intermediate products which have been shown to cause oxidative damage to the target tissues such as the liver and the lung [45]. Strenuous physical exercise in air-polluted environments exposes the athlete to higher concentrations of the oxidative pollutants over a shorter period of time compared to the sedentary individual. Outdoor exercise programs which result in the exposure of greater than normal levels of the skin to ultraviolet light (UVA and UVB) may also contribute to an increased burden of reactive oxygen species in the athlete as compared to the sedentary person.

Measurement of Free Radical Activity

The generation of free radicals during endurance exercise has been directly quantified in muscle tissue by the use of electron paramagnetic resonance (EPR) [10, 21] and indirectly by trapping very short-lived radi-

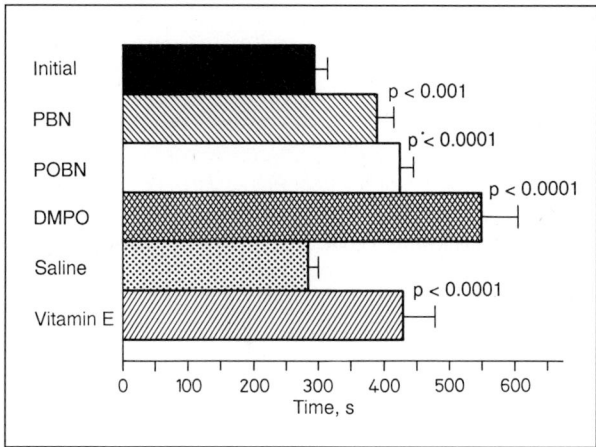

Fig. 2. Effect of free radical spin trappers and vitamin E on the endurance of mice. Adapted from Novelli et al. [34].

Table 1. Indices of free radical activity

Breath	Ethane, pentane
Muscle	Electron paramagnetic and spin resonance activities
Serum/plasma	Malondialdehyde, lipid peroxides, conjugated dienes, uric acid, mitochondrial-GOT, β-glucuronidase, pyruvate kinase, creatine kinase
Red blood cells	Hemolysis, malondialdehyde, lipid peroxides
Urine	Lipid peroxides, DNA oxidation products, malondialdehyde

cals with compounds called spin trappers. The adduct formed can be measured using electron spin resonance spectroscopy. Recently, the oral administration of electron spin trappers resulted in an increase in endurance exercise which is linked to the consequent decrease in in vivo free radicals (fig. 2) [34] (see section on vitamin E and exercise).

The products of free radical attack on muscle fiber lipids include lipid hydroperoxides and aldehydes, such as malondialdehyde [10], volatile fatty acid oxidation products including ethane and pentane [11, 13], lipid alcohols, conjugated dienes [27, 28] and epoxy-fatty acids (table 1) [45].

Pentane (a breakdown product of n-6 fatty acids, such as arachidonic acid) exhalation in the breath is increased following exericse [11] as are

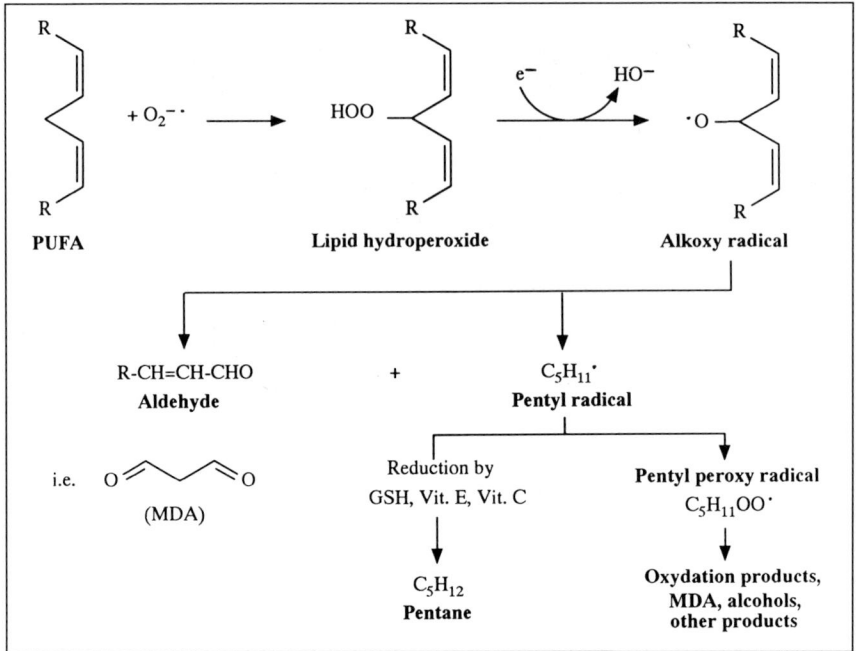

Fig. 3. Major products of lipid peroxidation, including malondialdehyde (MDA) and pentane. Adapted from Pryor [52].

serum levels of malondialdehyde and conjugated dienes (fig. 3). Traditionally, serum (or plasma) malondialdehyde levels have been indirectly determined by using the thiobarbituric acid reactive substances (TBAR) assay, which is not specific for malondialdehyde [49]. More recently, malondialdehyde has been measured directly, using HPLC [48] although many investigators continue to use the TBAR assay. Both pentane and malondialdehyde are intermediate products of lipid peroxidation and both can be metabolized in vivo [44, 48]. Thus, increased levels of these compounds may represent an overwhelming of the normal metabolic control mechanisms which ordinarily maintain a low concentration of these compounds in vivo. For example, life-long exercise training of rats resulted in a 21% increase in TBAR compounds from gastrocnemius homogenates which was significantly greater than found in muscle from sedentary rats [46]. In contrast, Alessio and Goldfarb [1] found that red and white muscle from sedentary

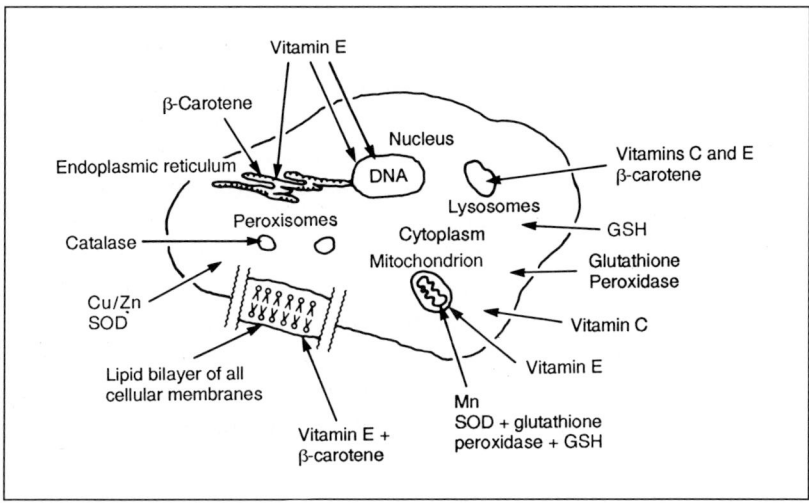

Fig. 4. Cellular antioxidant-protective compounds. Adapted from Machlin and Bendich [31].

rats had significantly higher concentrations of TBARs which were further elevated by exercise compared to muscles from rats trained for 18 weeks.

Indirect indices of muscle membrane leakage include the increased concentration of the muscle enzyme, pyruvate kinase in the serum. Recently, the increase in pyruvate kinase has been directly correlated with increased red blood cell hemolysis (another index of membrane susceptibility to lipid peroxidation) and liver malondialdehyde levels as well as with decreased concentrations of glutathione peroxidase [9].

Antioxidant Defenses

In order to stabilize itself, the free radical abstracts an electron from a stable compound which, in turn, is transformed into a new free radical. This chain reaction continues until the free radical containing the lone electron pairs up with another molecule with an unpaired electron, or is deactivated by a chain reaction-breaking antioxidant. There are also certain free radical scavengers and enzymes which can facilitate the decomposition of reactive molecules which are precursors of free radicals. These antioxidants can break the chain reaction and/or lower the free radical burden before a chain reaction begins (fig. 4) [31].

Antioxidant Enzymes

Two important antioxidants are metalloenzymes which can interfere with the production of free radicals during the initiation phase by inactivating precursor molecules. There are two types of superoxide dismutases; a Mn-containing enzyme in mitochondria, and a Cu/Zn-containing enzyme in the cytoplasm, both of which catalyze the reduction of superoxide free radicals to hydrogen peroxide. Catalase, an Fe-containing enzyme found in peroxisomes, catalyzes the decomposition of the hydrogen peroxide produced as a result of superoxide dismutation or by other reactions. The end products of hydrogen peroxide breakdown are oxygen and water. These enzymes lower the potential for the formation of the highly reactive hydroxyl radical as well as other energized, oxygen-containing species. Selenium is an essential component of glutathione peroxidase. This enzyme is important in the decomposition of hydrogen peroxide and lipid peroxides [41].

Cardiac muscle has significantly lower levels of catalase ($<2\%$ of the activity found in liver) and superoxide dismutase ($<50\%$ of the activity found in liver) but has similar activity of glutathione peroxidase as seen in liver [12]. Skeletal muscle has $<5\%$ of the glutathione reductase activity as liver and a similar relationship is found for cytosolic levels of glutathione peroxidase and reductase, catalase, glucose-6-phosphate dehydrogenase and Cu/Zn superoxide dismutase [38]. Recently, Lew [26] has examined the effect of endurance training on the activities of these antioxidant enzymes in rat skeletal muscle. All four enzyme activities were significantly enhanced following training (from 33 to 74%) as compared to that seen in muscle from untrained animals whereas liver enzyme activities remained unchanged. These data are in partial agreement with the findings of Alessio and Goldfarb [1] who compared the levels of TBARs in red and white muscle in sedentary and trained rats before and after exercise. Sedentary rats had significantly higher TBARs in both muscles before and following exercise compared to trained rats; exercise increased lipid peroxidation levels regardless of training. Catalase activity in red muscle was significantly lower in trained versus sedentary muscle prior to exercise. Following exercise, catalase activity was decreased in red and white muscle from sedentary animals. In contrast, there was almost in 90% increase in catalase activity in muscles from trained rats following exercise. However, no significant change in superoxide dismutase activity was seen. There are other conflicting reports in the literature on the effects of exercise training on antioxidant enzymes [reviewed in 46].

Antioxidant Vitamins

Vitamins C and E and β-carotene (provitamin A) can directly interfere with the propagation stage of free radical generation and scavenge free radicals. Vitamin E (α-tocopherol), the major lipid-soluble antioxidant present in all cellular membranes, protects against lipid peroxidation [31]. Humans and rats have approximately the same concentration of vitamin E in skeletal muscle as in the liver; however, upon depletion, the liver loses its vitamin E very rapidly whereas muscle retains the vitamin [6, 29].

Vitamin C (ascorbic acid), is water-soluble and along with vitamin E, can quench free radicals as well as singlet oxygen. Ascorbate can also regenerate the reduced, antioxidant form of vitamin E [3]. The ascorbic acid content of skeletal muscle in human subjects is approximatley one third the level found in liver (3–4 mg/100 g in muscle and 10–16 mg/100 g wet tissue in liver [19]).

β-Carotene is an efficient quencher of singlet oxygen and can function as an antioxidant [8]. β-Carotene is the major carotenoid precursor of vitamin A. Vitamin A, however, cannot quench singlet oxygen and has a very small capacity to scavenge free radicals [5]. In rats fed diets containing 200 mg/kg β-carotene, the liver concentration was approximately 50 μg/g tissue compared to 0.03 μg/g in skeletal muscle. As with vitamin E, muscle retains β-carotene to a greater degree than liver during depletion; the time for 50% turnover was 9 days for liver and 18 days for muscle [40]. The concentration of β-carotene in human liver is approximately 1μg/g compared to 0.1 μg/g in the heart; skeletal muscle levels have not been reported [23].

Other Antioxidants

In addition to the vitamins and mineral-containing enzymes, there is a growing list of compounds with important antioxidant potentials [2]. Two of these compounds, glutathione (a tripeptide composed of glutamine, cysteine and glycine) and carnosine (a dipeptide composed of alanine and histidine) are especially important for the protection of skeletal muscle from oxidative damage [24, 32].

Glutathione is a critical source of reducing equivalents and cosubstrate for glutathione peroxidase. Glutathione is involved in inactivation of oxygen radicals as well as the regeneration of certain antioxidants, such as vitamin E [15]. Deprivation of glutathione from skeletal muscle results in muscle fiber necrosis and infiltration of phagocytic white blood cells. Mitochondrial swelling and membrane damage are evident on electromicrographs of skeletal muscle (but not cardiac muscle) from mice given buthionine sulfoximine, an inhibitor of glutathione synthesis. Moreover, mitochondria cannot synthesize glutathione de novo and thus any oxidative damage to mitochondrial membranes may result in an inability to transfer glutathione from the cytoplasm into the mitochondria [32].

Carnosine has recently been shown to scavenge peroxyl radicals, quench singlet oxygen, act as a reducing agent and is a critical chelator of copper. Human muscle contains a high concentration of carnosine (2–20 mM) as well as one third the total body pool of copper. Copper, as with other transition metals such as iron, can participate in chemical reactions which result in the formation of free radicals. The antioxidant properties as well as the chelation of copper by carnosine can help to protect muscle tissue from oxidative damage [24].

Vitamin E and Skeletal Muscle

Vitamin E is essential for the maintenance of skeletal muscle integrity. The requirement was first described in 1928 by Evans and Burr [reviewed in 33] in vitamin E-deficient rats with progressive hind limb paralysis. Histological examination of the paralyzed muscle showed severe muscle necrosis, with changes in the cross-sectional diameters of the muscle fibers. Phagocytic cells, attracted to the necrotic tissue, invade the damaged areas resulting in sites of inflammation. Electron micrographs show disrupted mitochondria and evidence of lysosomal enzyme activity. There is an increase in oxygen consumption in vitamin E-deficient skeletal muscle which may reflect an uncoupling of mitochondrial oxidative pathways.

The mitochondrial inner membrane, the site of ATP production, contains a ratio of vitamin E:phospholipids of 1:200. The importance of vitamin E in the functioning of mitochondria is clearly seen under conditions of deficiency. The mitochondria from vitamin E-deficient skeletal muscle have been carefully characterized by Quintanilha and Packer [38] and Davies et al. [10]. There is an increase in the free radical signal; lipid peroxidation (TBAR and conjugated dienes); and the negative inner membrane surface charge. Decreases in transmembrane potential, respiratory coupling and electron transport were also shown. The exercise endurance of the vitamin E-deficient rats was decreased by approximately 35%.

The vitamin E content of subcellular fractions of rabbit skeletal muscles shows that the sarcoplasmic reticulum from red (slow) muscle fibers

Table 2. Vitamin E content of rabbit muscle

	Muscle vitamin E[1]	
	fast (white)	slow (red)
Homogenate	1,530	2,859
Mitochondria	3,375	4,208
Sarcoplasmic reticulum	9,760	43,440

Adapted from Salviati et al. [39].
[1] α-[^{3}H]-tocopherol (dpm/mg protein).

has approximately a 4-fold higher concentration of vitamin E than found in fast white fibers (table 2) [39]. Fast fibers, when stimulated, fatigue faster and release more enzymes into the serum and have a greater susceptibility to necrosis associated with vitamin E deficiency.

The concentration of vitamin E required in the diet to prevent skeletal muscle damage can be assessed by examining the serum pyruvate kinase concentration, an enzyme released specifically from damaged skeletal muscle [30]. Serum pyruvate kinase levels are normalized in rats consuming diets containing 15 mg/kg but not 7 mg/kg vitamin E, which is sufficient to prevent testes degeneration [4]. Red blood cell hemolysis is correlated with peroxidative damage and serum pyruvate kinase levels [9]. Peroxidative hemolysis of red blood cells was eliminated when rat diets contained 50 mg/kg (the standard level found in rat diets), yet 15 mg/kg was sufficient to normalize skeletal muscle enzyme leakage [4].

Effect of Exercise on Vitamin E Status

Laboratory Animals Fed Vitamin E-Sufficient Diets. Endurance training results in a significant increase in mitochondrial content of skeletal muscles. There is also a concomitant increase in the mitochondrial respiratory activity. Gohil et al. [16] showed almost a doubling of the mitochondrial succinate cytochrome c reductase activity in muscle from trained compared to sedentary rats. The vitamin E content of the muscle, however, remained constant, suggesting that the mitochondria from the trained rats may be under oxidative risk. There is an overall decrease in the vitamin E: oxidative capacity ratio as reflected in an increase in ubiquinone content, an indicator of oxidation (table 3).

Table 3. Endurance training: effect on mitochondrial activity and vitamin E levels in rat muscle

Muscle	Sedentary	Trained
Mitochondrial succinate cytochrome c reductase, µmol/min/g	3.6	7.1*
Vitamin E red quadriceps, nmol/g	29.3	29.0

Adapted from Gohil et al. [16].
* $p < 0.05$.

Laboratory Animals Fed Vitamin E-Deficient Diets. The effects of vitamin E deficiency on oxidative indicators in exercised animals was first examined by Brady et al. [7]. Rats swum to exhaustion had a significant increase in muscle TBAR compounds compared to sedentary rats. Vitamin E deficiency further elevated the concentration of TBAR in muscle, which remained elevated for the next 24 h. Davies et al. [10] determined the concentration of free radicals (using EPR spectroscopy) in muscles from rats given a vitamin E-deficient diet or a diet containing 21 mg/kg vitamin E for 6 months (approximately half the standard level of 50 mg/kg). There was a greater concentration of radicals in the deficient muscle, which was further increased by exercise. The concentration of radicals in the muscle from the rats fed 21 mg/kg, however, increased to a greater degree than seen in the deficient muscle following exercise. The vitamin E-deficient rats had significant red blood cell hemolysis at 100 days on the deficient diet. Maximal endurance was 28 min in the deficient group compared to 46 min in the controls.

In another experiment from the Berkeley group, deficient rats were given 10 mg/kg body weight and the control group was given 40 mg/kg body weight of vitamin E. Deficiency was associated with a 75% reduction in endurance. In addition, the concentration of vitamin E in the muscle was significantly decreased from 7.2 to 4.9 µg/g following exercise in the control rats. In contrast to vitamin E, total plasma glutathione levels are increased during exercise training although the percentage of oxidized glutathione is increased in the tissue [36].

Pentane exhalation is significantly increased in vitamin E-deficient rats, and approximately 11 mg/kg diet is sufficient to return resting pentane levels to those seen in rats fed diets containing 30–50 mg/kg. Exercise

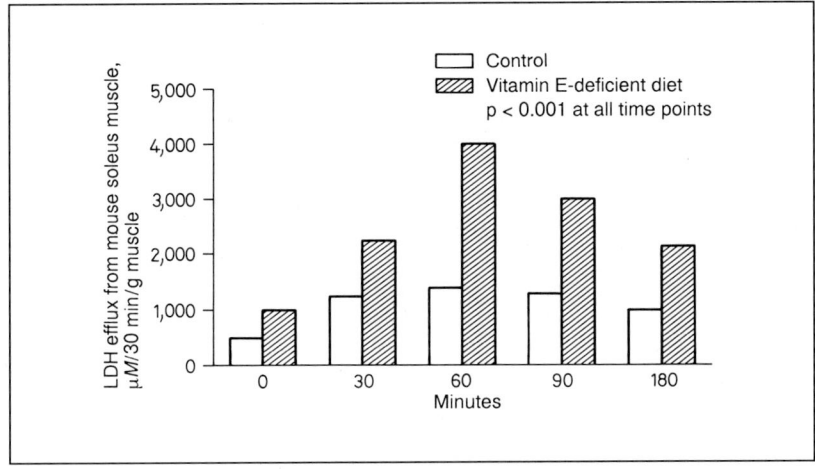

Fig. 5. Muscle enzyme leakage associated with skeletal muscle contraction. Adapted from Jackson et al. [22].

increases pentane exhalation in the rats given vitamin E, but does not further increase exhalation in the deficient animals [13].

Laboratory Animals Fed Vitamin E-Supplemented Diets. Jackson et al. [22] have used electrical stimulation of skeletal muscle to observe the effects of repetitive contractions in anesthetized rats. Responses of muscles from vitamin E-deficient rats were compared to those seen in rats fed a supplemental dietary level of 240 mg/kg. Damage to muscle membranes was inferred from the 150% increase in plasma creatine kinase from deficient rats following the stimulation. There was no change in this plasma enzyme level in the supplemented group. The concentration of lactate dehydrogenase released from the deficient muscle was increased 4-fold compared to a 2-fold increase seen in the supplemented muscle (fig. 5).

Enzyme release following contractions has been associated with depletion of ATP stores and an influx of extracellular calcium into the muscle fiber. Calcium influx can initiate membrane enzymes, such as PLA_2, resulting in the release of arachidonic acid from the membrane. The arachidonic acid cascade of oxidative reactions leads to the possible formation of prostaglandins (PGE_2) and leukotrienes (mentioned in the Free Radicals section). Vitamin E has been shown to interfere with PLA_2

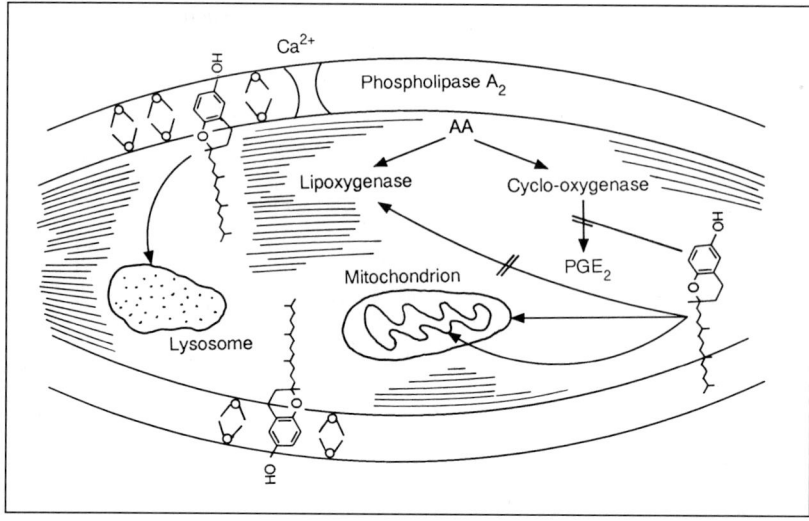

Fig. 6. Potential protective actions of vitamin E in skeletal muscle. The vitamin E molecule is increased in size manyfold to permit the structural formula to be shown. Vitamin E is approximately the same size as a fatty acid molecule.

activity and inhibit lipoxygenase activity and PGE_2 production [reviewed in 30].

Vitamin E also protects muscle membranes from oxidative damage (fig. 6). Muscle malondialdehyde levels in vivo were not significantly different between groups, however, when muscle tissue from deficient rats following stimulation was incubated in an oxidizing medium, the concentration of malondialdehyde was 2-fold that found in the supplemented group. Lysosomal, mitochondrial as well as sarcoplasmic membranes are potential sites of oxidative damage. The muscle tissue from the supplemented group had approximately 3 times the concentration of vitamin E found in the deficient muscle (23.6 and 7.3 µg/g respectively) [21, 22].

In a novel protocol, Novelli et al. [34] measured the maximal swimming endurance in mice given injections of electron spin trappers or 100 mg/kg vitamin E (fig. 2). The hypothesis is that the electron spin trappers and the vitamin E lower the free radical burden increased during strenuous exercise. The authors suggest that free radicals are involved in the regulation of muscle responses to endurance exercise which leads to fatigue.

Table 4. Effect of acute intense exercise on serum and red blood cell vitamin E

Measurements	Lactate	Vitamin E	
		Plasma, µg/ml	RBC, µg/ml
Baseline	1.3	8.4	1.8
During exercise	11.1*	10.6*	3.2*
End of exercise	13.7*	9.6*	2.5*
10 min postexercise	12.2*	8.4	1.7

Adapted from Pincemail et al. [37].
* Significantly different when compared to baseline value.

Effect of Vitamin E on Exercise

Humans Eating Standard Diets: No Supplementation. Pincemail et al. [37] examined the effects of acute, intensive exercise (ergometer bicycling to exhaustion) on plasma lactic acid, vitamin E and red blood cell vitamin E levels in adult males (presumably untrained). Lactic acid levels increased 10-fold above baseline at the end of the exercise, indicating anaerobic metabolism. Plasma and red blood cell vitamin E levels rose significantly during the first few minutes of exercise and returned to normal following a 10-min postexercise rest period (table 4). The mobilization of vitamin E may be indirect and reflect the release of fatty acids from adipose tissue, which contains high concentrations of vitamin E. The lack of a finding of a decrease in vitamin E status following exercise as shown in animal studies [36] may be due to the length of the exercise period. Ohno et al. [35] showed that short-term, acute exercise in sedentary young men did not affect a number of indices of red blood cell oxidative stress.

Vitamin E Supplementation. Young, untrained males were exercised to exhaustion before and following supplementation for 4 weeks with 300 mg vitamin E [47]. The vitamin E supplementation did not affect exercise time or other indices of endurance. Vitamin E supplementation did lower the pre-exercise, baseline level of serum TBAR and β-glucuronidase (an index of cellular damage). Following exercise, lipid peroxides were significantly increased in the control group, but were decreased significantly in the vitamin E-supplemented group. Serum uric acid levels were increased in both groups, as were mitochondrial-GOT and β-gluc-

Table 5. Vitamin E supplementation (300 mg/day/4 weeks) and its effects on antioxidant status following exhaustive exercise

Serum	Group	Exercise	
		pre	post
Lipid peroxides (TBAR), nmol/ml	control	4.4[a]	4.5[c]
	vitamin E	3.8[b]	3.5[d]
Uric acid, mg/dl	control	6.0[a]	8.7[b]
	vitamin E	5.7[a]	8.0[b]
Mitochondrial-GOT, U	control	3.4[a]	5.6[b]
	vitamin E	2.8[a]	4.1[b]
β-Glucuronidase, U	control	1,274[a]	1,510[c]
	vitamin E	915[b]	964[b]

Adapted from Sumida et al. [47].
[a-d] Differences in letters in groups between columns and/or rows are significant.

uronidase, although the relative increases were lower in the vitamin E group (table 5). In a separate study, also involving supplementation with 300 mg/day vitamin E (for 6 weeks), compared to a placebo control group of untrained young males, strenuous exercise resulted in a significant rise in serum creatine kinase, which was similar in both groups [17]. These data differ from the findings of Jackson et al. [22] in rats. Trained athletes in the vitamin E or placebo group participating in this protocol did not demonstrate an increase in serum creatine kinase following exercise.

Exercise also increases the concentration of breath pentane. Supplementation of adult males with 1,200 mg/day of vitamin E for 2 weeks lowered the baseline pentane exhalation as well as lowering the level exhaled following exercise [11]. Exercise training increases the endurance of athletes. Mountain climbers add an additional oxidative stress to exercise performed at high altitude because of the relative hypoxic conditions. Simon-Schnass and Pabst [43] investigated the effects of exhaustive bicycle ergometer exercise in trained mountain climbers at 2,500 and 5,000 m. All 12 athletes were given a multivitamin-mineral supplement daily throughout the protocol. The athletes were randomly placed in two groups: placebo or 400 mg vitamin E/day for 4 weeks. Pentane exhalation increased significantly in the placebo group following exercise, but did not

Table 6. Vitamin E supplementation (400 mg/day/4 weeks) and effects on antioxidant status following exhaustive exercise at high altitudes

Measurement	Group	Exercise	
		pre	post
Pentane exhalation, ppm	placebo	0.88[a]	1.48[b]
	vitamin E	0.92[a]	0.97[a]
Anaerobic threshold, W	placebo	231[a]	226[a]
	vitamin E	242[a]	293[b]

Adapted from Simon-Schnass and Pabst [43].
[a,b] Differences in letters in groups are significant.

increase in the vitamin E-supplemented group. The anaerobic threshold, defined as the level of exercise performed prior to an increase in serum lactic acid, was significantly increased in the vitamin E group following exercise and remained unchanged in the placebo group (table 6). Exhaustive exercise did not affect serum uric acid levels or serum enzyme level measured, which may reflect the training of the athletes and/or the higher level of vitamin E supplementation in this study compared to the subjects examined by Sumida et al. [47].

Conclusion

Exercise has been associated with increases in free radical production in skeletal muscle and can be detected systemically. The consequences of the overproduction of the oxidative species include damage to mitochondria and sarcoplasm. Thus the antioxidant protection which is sufficient under sedentary conditions is overwhelmed by acute, strenuous physical activity, even in the healthy young male. Deficiency in vitamin E, the major lipid-soluble antioxidant present in tissues, further exacerbates the oxidative damage associated with exercise. Endurance training and vitamin E supplementation help to restore the free radical:antioxidant balance and reduce the levels of sarcoplasmic membrane leakage and indices of lipid peroxidation. Current research indicates that antioxidant requirements are increased by strenuous exercise.

References

1. Alessio HM, Goldfarb AH: Lipid peroxidation and scavenger enzymes during exercise: adaptive response to training. J Appl Physiol 1988;64:1333–1336.
2. Ames BN: Endogenous oxidative DNA damage, aging, and cancer. Free Radic Res Commun 1989;7:121–128.
3. Bendich A, Machlin LJ, Scandurra O, et al: The antioxidant role of vitamin C. Free Radic Biol Med 1986;2:419–444.
4. Bendich A, Gabriel E, Machlin LJ: Dietary vitamin E requirement for optimum immune responses in the rat. J Nutr 1986;116:675–681.
5. Bendich A, Olson JA: Biological actions of carotenoids. FASEB J 1989;3:1927–1932.
6. Bieri JG: Kinetics of tissue alpha-tocopherol depletion and repletion; in Nair PP, Kayden HJ (eds): Vitamin E and Its Role in Cellular Metabolism. IV. Aspects of Vitamin E Metabolism Relating to Dietary Requirement. New York, The New York Academy of Sciences, 1972, vol 203.
7. Brady PS, Brady LJ, Ullrey DE: Selenium, vitamin E and the response to swimming stress in the rat. J Nutr 1979;109:1103–1109.
8. Burton GW, Ingold KV: Beta-carotene: an unusual type of lipid antioxidant. Science 1984;224:569–573.
9. Chow CK: Effect of dietary vitamin E and selenium on rats: Pyruvate kinase, glutathione peroxidase and oxidative damage. Nutr Res 1990;10:183–194.
10. Davies KJA, Quintanilha AT, Brooks GA, Packer L: Free radicals and tissue damage produced by exercise. Biochem Biophys Res Commun 1982;107:1198–1205.
11. Dillard CJ, Litov RE, Savin WM, Dumelin EE, Tappel AL: Effects of exercise, vitamin E, and ozone on pulmonary function on lipid peroxidation. J Appl Physiol 1978;45:927–932.
12. Doroshow JH, Locker GY, Myers CE: Enzymatic defenses of the mouse heart against reactive oxygen metabolites. J Clin Invest 1980;65:128–135.
13. Gee DL, Tappel AL: The effect of exhaustion on expired pentane as a measure of in vivo lipid peroxidation in the rat. Life Sci 1981;28:2425–2429.
14. Gerster H: Review: The role of vitamin C in athletic performance. J Am Coll Nutr 1989;8:636–643.
15. Gibson DD, Hawrylko J, McCay PB: GSH-dependent inhibition of lipid peroxidation: Properties of a potent cytosolic system which protects cell membranes. Lipids 1985;20:704–711.
16. Gohil K, Rothfuss L, Lang J, Packer L: Effect of exericse training on tissue vitamin E and ubiquinone content. J Appl Physiol 1987;63:1638–1641.
17. Helgheim I, Hetland O, Nilsson S, Ingjer F, Stromme SB: The effects of vitamin E on serum enzyme levels following heavy exercise. Eur J Appl Physiol 1979;40:283–289.
18. Holloszy JO, Booth FW: Biochemical adaptations to endurance exercise in muscle. Annu Rev Physiol 1976;38:273–291.
19. Hornig D: Distribution of ascorbic acid, metabolites and analogues in man and animals; in King CG, Burns JJ (eds): Second Conference on Vitamin C. New York, The New York Academy of Sciences, 1975, vol 258, pp 109–118.
20. Jackson MJ: Free radicals and skeletal muscle disorders; in Das DK, Essman WB

(eds): Oxygen Radicals: Systemic Events and Disease Processes. New York, Karger, 1990, chapt 6, pp 149–171.
21 Jackson MJ, Edwards RHT, Symons MCR: Electron spin resonance studies of intact mammalian skeletal muscle. Biochim Biophys Acta 1985;847:185–190.
22 Jackson MJ, Jones DA, Edwards RHT: Vitamin E and skeletal muscle; in Biology of Vitamin E. Ciba Foundation Symposium 101. London, Pitman Press, 1983, pp 224–239.
23 Kaplan LA, Lau JM, Stein EA: Carotenoid composition, concentrations, and relationships in various human organs. Clin Physiol Biochem 1990;8:1–10.
24 Kohen R, Yamamoto Y, Cundy KC, Ames BN: Antioxidant activity of carnosine, homocarnosine, and anserine present in muscle and brain. Proc Natl Acad Sci USA 1988;85:3175–3179.
25 Korthius R, Granger N, Foronsky M, Tayler A: The role of oxygen derived free radicals in ischemia induced increases in canine skeletal muscle vascular permeability. Circ Res 1985;57:599.
26 Lew H: The effects of physical exercise on tissue antioxidative capacity: The role of glutathione. Berkeley, University of California at Berkeley, 1989.
27 Lindsay T, Romaschin A, Walker PM: Free radical mediated damage in skeletal muscle. Microcirc Endothelium Lymphatics 1989;5:157–169.
28 Lindsay T, Walker P, Mickle D, Romaschin A: Measurement of hydroxy conjugated dienes after ischemia/reperfusion in canine skeletal muscle. Am J Physiol 1988;254: H578–H583.
29 Machlin LJ: Vitamin E; in Machlin LJ (ed): Handbook of Vitamins – Nutritional, Biochemical and Clinical Aspects. New York, Dekker, 1984, pp 99–146.
30 Machlin LJ: Vitamin E; in Machlin LJ (ed): Handbook of Vitamins – Nutritional, Biochemical and Clinical Aspects. New York, Dekker, 1990, pp 99–145.
31 Machlin LJ, Bendich A: Free radical tissue damage: protective role of antioxidant nutrients. FASEB J 1987;1:441–445.
32 Martensson J, Meister A: Mitochondrial damage in muscle occurs after marked depletion of glutathione and is prevented by giving glutathione monoester. Proc Natl Acad Sci USA 1989;86:471–475.
33 Nelson JS: Pathology of vitamin E deficiency; in Machlin LJ (ed): Vitamin E – A Comprehensive Treatise. Basic and Clinical Nutrition. New York, Dekker, 1980, vol 1, pp 397–428.
34 Novelli GP, Bracciotti G, Falsini S: Spin-trappers and vitamin E prolong endurance to muscle fatigue in mice. Free Radic Biol Med 1990;8:9–13.
35 Ohno H, Sato Y, Yamashita K, Doi R, Arai K, Kondo T, Taniguchi N: The effect of brief physical exercise on free radical scavenging enzyme systems in human red blood cells. Can J Physiol Pharmacol 1986;64:1263–1265.
36 Packer L: Vitamin E, physical exercise and tissue damage in animals. Med Biol 1984;62:105–109.
37 Pincemail J, Deby C, Damus G, Pirnay F, Bouchez R, Massaux L, Goutier R: Tocopherol mobilization during intense exercise. Eur J Appl Physiol 1988;57:189–191.
38 Quintanilha AT, Packer L: Vitamin E, physical exercise and tissue oxidative damage; in Porter R, Whelan J (eds): Biology of Vitamin E. Ciba Foundation Symposium 101. London, Pitman Press, 1983, pp 56–69.

39 Salviati G, Betto R, Margreth A, Novello F, Bonetti E: Differential binding of vitamin E to sarcoplasmic reticulum from fast and slow muscles of the rabbit. Experientia 1980;36:1140–1141.
40 Shapiro SS, Mott DJ, Machlin LJ: Kinetic characteristics of beta-carotene uptake and depletion in rat tissue. J Nutr 1984;114:1924–1933.
41 Sies H (ed): Oxidative Stress. New York, Academic Press, 1985.
42 Simic MG, Bergtold DS, Karam LR: Generation of oxy radicals in biosystems. Mutat Res 1989;214:3–12.
43 Simon-Schnass I, Pabst H: Influence of vitamin E on physical performance. Int J Vitam Nutr Res 1988;58:49–54.
44 Siu GM, Draper HH: Metabolism of malonaldehyde in vivo and in vitro. Lipids 1982;17:349–355.
45 Slater TF: Free radical-mediated tissue damage. Nutrition 1987;1:46–50.
46 Starnes JW, Cantu G, Farrar RP, Kehrer JP: Skeletal muscle lipid peroxidation in exercised and food-restricted rats during aging. J Appl Physiol 1989;67:69–75.
47 Sumida S, Tanaka K, Kitao H, Nakadomo F: Exercise-induced lipid peroxidation and leakage of enzymes before and after vitamin E supplementation. Int J Biochem 1989;21:835–838.
48 Wade CR, van Rij AM: In vivo lipid peroxidation in man as measured by the respiratory excretion of ethane, pentane and other low-molecular-weight hydrocarbons. Anal Biochem 1985;150:1–7.
49 Wade CR, van Rij AM: Plasma thiobarbituric acid reactivity: reaction conditions and the role of iron, antioxidants and lipid peroxy radicals on the quantitation of plasma lipid peroxides. Life Sci 1988;43:1085–1093.
50 Williams MH: The role of vitamins in physical activity; in Nutritional Aspects of Human Physical and Athletic Performance. Illinois, Thomas, 1985, pp 147–185.
51 Williams MH: Vitamin, iron and calcium supplementation: Effect on human physical performance; in Haskell W, Scala J, Whittam J (eds): Nutrition and Athletic Performance. Proc Conf on Nutritional Determinants in Athletic Performance, 1981, pp 106–153.
52 Pryor WA; Can vitamin E protect humans against pathological effects of ozone in smog? J Nutr 1991;53:702–722.

Adrianne Bendich, PhD, Senior Clinical Research Coordinator,
Clinical Nutrition, Hoffmann-La Roche Inc., Nutley, NJ 07110 (USA)

A Biochemical Mechanism to Explain Some Characteristics of Overtraining

E.A. Newsholme[a], M. Parry-Billings[a], N. McAndrew[a], R. Budgett[b,1]

[a] Department of Biochemistry, University of Oxford, Oxford, and
[b] British Olympic Medical Centre, Northwick Park Hospital, Harrow, Middx, UK

The Overtraining Syndrome

A syndrome is a set of symptoms which occur together, usually, it is assumed, as a consequence of one underlying cause. The term is legitimately applied to overtraining since decreased athletic performance is just one of many signs and symptoms (see below).

Exercise elicits a 'stress' response and results in a degree of short-term fatigue, from which the athlete recovers. If exercise is performed on a regular basis, 'adaptation' occurs and the athlete's performance improves. This is the *training* effect. However, if exercise periods are too frequent, too intense and/or too prolonged, possibly combined with inadequate nutrition, recovery after each exercise bout is not complete, less adaptation and hence less improvement in performance occurs. If continued, this can lead to overtraining [15]. The overtrained athlete may be said to be suffering from *the overtraining syndrome*.

The overtraining syndrome is a complex of clinical conditions, which has been described by a number of authors. And it has been described not only in human athletes but also in racehorses. The mechanism responsible for the overtraining syndrome in athletes may be involved in other condi-

---

[1] We wish to thank Professor C. Williams and Dr. S. Brooks, Department of Physical Education and Sports Sciences, University of Loughborough, for providing the plasma blood samples for some of the results reported in table 5.

tions of stress; some of the symptoms of athletic overtraining have been reported, for example, in stressed businessmen or students taking exams.

The overtraining syndrome can be caused by too much exercise: it can also be caused by a combination of too much exercise with some other form of stress [8]. There is considerable individual variability in regards to the development of overtraining. Verma et al. [23] put a group of 15 male subjects through a common training programme with the result that 10 subjects improved and 5 became overtrained. What determines the variability in susceptibility to this condition is not known. There is, however, a possible common mechanism to account for many of the signs and symptoms of overtraining which might lead to an explanation for this variability. There may also be a simple biochemical explanation for the effects of overtraining on the immune system. These are discussed below.

The Imbalanced Amino Acid Hypothesis to Explain Some Signs and Symptoms of the Overtrained State

It is proposed that *excessive* exercise can alter chronically the balance of the plasma concentration of branched-chain amino acids and free tryptophan which increases *chronically* the level of 5-hydroxytryptamine in the brain and peripheral nerve cells. The mechanism(s) underlying these chronic changes is not known, but it is possible to speculate that there is a prolonged increase in the plasma concentration of free tryptophan possibly due to a chronic elevation in the plasma level of long chain fatty acids.

Amino Acids and Exercise

Amino acids are the building blocks from which proteins are made via the process known as protein synthesis. The significance of this latter process, which has been emphasized over recent years, may have been responsible for a decrease in interest in other roles of amino acids. One important role of some amino acids is that they act as precursors for certain brain neurotransmitters – the chemicals that carry out specific messenger roles in the brain. One of these amino acids is tryptophan, which is converted in the brain to a neurotransmitter known as 5-hydroxytryptamine (5-HT). There is evidence that tiredness and sleep may be, in part, influenced by the brain level of 5-HT; it is possible, therefore, that this neurotransmitter may be involved in fatigue, that is, fatigue originating in the brain, known as central fatigue.

Table 1. Plasma concentrations of fatty acids, branched-chain amino acids, and tryptophan before and after the Stockholm marathon

Condition	Plasma concentrations µM or mM		Total tryptophan	Free tryptophan	Free tryptophan/ branched chain
	fatty acid	branched-acid amino acid			
Pre-run	350	470	55	7.7	1.6
Post-run	1,550*	380*	57	18.0*	4.7

* $p < 0.001$ as compared with the pre-exercise value (Student's t test). Data from Blomstrand et al. [4].

There are two important additional facts. First, branched-chain amino acids (leucine, isoleucine and valine) are *not* taken up by liver but are taken up by muscle where they are used, primarily, for energy formation. And their rate of oxidation is increased during exercise [24]. Secondly, both branched-chain amino acids and tryptophan enter the brain upon the same amino acid carrier. Hence there is competition between the two types of amino acids for entry into brain.

A decrease in the level of branched-chain amino acids in the blood, due to an increase rate of utilization by muscle, will increase the ratio of tryptophan to branched-chain amino acids in the bloodstream and favour the entry of tryptophan into the brain. This will increase the rate of formation of 5-HT and hence increase the level of this neurotransmitter in the brain. This could result in fatigue originating in the brain. A further important point is that the blood level of fatty acids may also play an additional role in this type of fatigue: an increase in the plasma fatty acid level above about 1 mM increases the free concentration of tryptophan – and it is probably the free concentration, rather than the total concentration of tryptophan, that influences the rate of entry of tryptophan into the brain. Hence an increase in the plasma fatty acid level plus a decrease in that of branched-chain amino acids could markedly influence the plasma concentration ratio, free tryptophan/branched-chain amino acids. And this has been shown to be the case in endurance activity (table 1) [4]. Furthermore, endurance exercise in rats produced similar changes in the free tryptophan/branched-chain amino acid concentration ratio in the plasma. How-

Table 2. Effect of sustained exercise in the rat on the concentrations (nmol g^{-1}) of tryptophan, 5-HT, 5-hydroxyindoleacetic acid (5-HIAA) in four areas of the brain

Brain area	Contents in brain, nmol/g					
	tryptophan		5-HT		5-HIAA	
	rested	post-ex	rested	post-ex	rested	post-ex
Cortex	23	34*	2.9	2.6	1.8	1.9
Cerebellum	23	30*	0.59	0.58	0.50	0.58
Hippocampus	25	35*	2.1	2.2	1.9	2.3*
Striatum	27	35*	2.3	2.6	2.5	3.2
Brain stem	24	32*	3.6	4.1*	2.7	3.2*
Hypothalamus	25	34*	4.9	5.7*	3.0	3.9*

Data are for sedentary and trained rats combined, and are presented as means: * $p < 0.05$ as compared with the resting value (Student's t test). For full details of methods and results, see Blomstrand et al. [5]. post-ex = post-exercise.

ever, more complete experiments are possible with animals and it has been shown that, when this plasma concentration ratio increases, the concentration of 5-HT in the hypothalamus and the brain stem increases (table 2) [5].

It seems likely that muscle will use branched-chain amino acids for energy formation when the glycogen level in the muscle becomes depleted. Similarly, it is possible that, as the muscle glycogen level approaches depletion, the blood level of fatty acids will increase above 1 mM and hence influence the plasma level of free tryptophan. Thus the fatigue that is caused by the depletion of glycogen in muscle could be due, not to a direct effect of glycogen depletion on the muscle, but to an increase in the level of 5-HT in a specific area of the brain. (Of course, the decrease in the glycogen level in the muscle will need to change some factor in muscle – possibly the activity of branched-chain keto acid dehydrogenase – that will initiate a series of events leading to a change in the 5-HT level in the brain.) Failure of the motor centre in the brain to stimulate muscle to contract would mean that the power output would have to fall – that is, fatigue due to a change in the balance of concentrations of key amino acids in the blood but achieved through an effect on the brain!

Table 3. Some signs and symptoms of the overtraining syndrome [data from 15]

Increased early morning heart rate	Increased incidence of infections
Retarded recovery of heart rate after exercise	Poor healing of wounds
	Loss of libido
Postural hypotension	Disturbed sleep
Increased resting blood pressure	Decreased appetite
Retarded recovery of blood pressure to basal after exercise	Depression
	Loss of drive and enthusiasm
Fall in haematocrit	Increased fluid intake at night
Decreased performance	Muscle and joint pains:
Amenorrhoea	heavy leggedness
Decreased V_{O_2max}	Increased serum creatine phosphokinase activity
Weight loss	

Such changes, especially if chronic, in the level of 5-HT in both the brain and peripheral nerves could cause physiological and behavioural changes other than fatigue.

5-HT and Overtraining Syndrome

5-HT-containing cells occur in several large clusters in the pons and upper medulla within groups of cells known as the raphe nuclei. Projections from these cells pass rostrally to most areas of the brain via the medial forebrain bundle: the cortex, hippocampus, limbic system and hypothalamus all receive 5-HT terminals [20]. Projections from the caudal nuclei run to the medulla and spinal cord. With such wide-ranging projections to areas involved in motor, neuroendocrine and 'emotive' functions, interpretation of changes in levels of 5-HT is difficult but prolonged changes in the level of this neurotransmitter in these areas of the brain could account for the wide-ranging effects of overtraining (table 3).

The physiological functions of 5-HT in the brain can be grouped into at least three areas: sleep, wakefulness and mood; motor neuron excitability; autonomic and endocrine function.

Sleep. Lesions in the raphe nuclei or the administration of *p*-chlorophenylalanine – an inhibitor of the synthesis of 5-HT, abolish sleep in animals; and the microinjection of 5-HT in specific medullary areas of the brain induces sleep. Could this neurotransmitter, therefore, also be involved in the fatigue associated with the overtrained state.

Motor Neuron Excitability. Descending 5-HT neurons increase motor neuron excitability and in doing so increase monosynaptic reflexes and decrease polysynaptic reflexes. Inhibition of polysynaptic reflexes may include those involved in exercise such as running and may contribute to the decreased maximal work capacity in the overtrained state.

Autonomic and Endocrine. From the medulla oblongata (brain stem) 5-HT neurons project to the hypothalamus which is considered to be the major centre for autonomic, endocrine and neuronal integration. Rang and Dale [20] observed that 5-HT inhibits the release of factors from the hypothalamus that act to control the rate of release of pituitary hormones. A low rate of gonadotrophin-releasing hormone (GnRH) secretion by the hypothalamus would be expected to decrease the rate of release of luteinizing hormone (LH) and follicle-stimulating hormone (FSH) from the pituitary and hence lower the plasma levels of these hormones. Nash [11] has reported a decrease in the LH pulse frequency and amplitude of LH secretion in overtrained man. This would be expected to lower the rate of testosterone synthesis and release with a subsequent decline in plasma levels of testosterone. In the female, such a decrease would be expected to interfere in the complex endocrine system that controls the menstrual cycle and could thus lead to irregular menses or amenorrhoea. Lightman and Everitt [9] provide some evidence to suggest that 5-HT plays a role in GnRH rhythmicity; the latter is inhibited by high levels of 5-HT; and if central levels of 5-HT are elevated pharmacologically, the pre-ovulatory LH surge is lost and amenorrhoea develops.

Lesions in the ventromedial hypothalamus, which decrease the level of 5-HT, lead to hyperphagia. Hence an elevation in the level of 5-HT in this area of the brain could explain the loss of appetite in overtrained subjects.

Peripheral 5-HT and Overtraining Syndrome

Similar considerations regarding entry of tryptophan and its conversion pathway to 5-HT apply to peripheral nerves as to central nerves [20]. Hence peripheral 5-HT levels could be elevated in the overtrained state. It is known to stimulate sympathetic afferent nerves in the heart which would cause an increase in heart rate, a well-established sign of overtraining. In contrast, it causes inhibition of the rate of noradrenaline release from sympathetic nerve endings on blood vessels; this would be expected to have a general vasodilatory effect and could explain changes in blood pressure in overtrained state.

Effects of Exercise on the Immune System

The evidence for an effect of exercise on the immune system is derived from two sources: laboratory-based investigations into specific aspects of immune function and epidemiological studies.

Exercise, particularly low-intensity exercise, appears to be beneficial for the immune system [6]. Thus, there is evidence that low-intensity exercise enhances the lymphocyte response to mitogenic stimulation in vitro and increases the number of natural killer cells: exercise may result in an increase in the number of circulating lymphocytes ('leukocytosis') [16]. These effects would be expected to enhance immune function [7].

In contrast, there is considerable evidence which suggests that exercise of high intensity and long duration is associated with *adverse* effects on immune function [6]. A period of training may prevent or reduce the magnitude of the post-exercise leukocytosis. Although the *total* number of lymphocytes may increase following exercise, changes in the numbers of lymphocytes in specific subpopulations may not enhance the overall potential immunological activity of the lymphocyte population; thus, the ratio of CD4[T_4] cells (helper):CD8[T_8] (suppressor) cells is decreased post-exercise [7, 10]. In addition to the effect of exercise on the *number* of circulating leukocytes, a decrease of immune *function* by exercise has also been reported. In general, high-intensity exercise and training appears to cause a marked *decline* in the functioning of cells of the immune system: the response of T lymphocytes to mitogenic stimulation in vitro may be decreased, antibody synthesis may be impaired [7] and immunoglobulin levels in blood and saliva are decreased post-exercise in trained subjects [21]; training reduces neutrophil and monocyte adherence and monocyte bacteriocidal activity [10], and training is associated with low levels of the complement components C3 and C4 [3].

The exercise-induced decrease in immune function appears not to be specific to a particular type of exercise. Indeed, it has been demonstrated in a large number of different types of athletes, including runners, swimmers, skiers and in ballet dancers. However, up until now, a mechanism to explain the decrease in immune function has not been put forward. We suggest that intense and long-duration exercise, particularly if it is regular, can cause a marked decrease in the plasma glutamine level and this can result in immunosuppression. In order to understand the evidence for this hypothesis, it is necessary to consider the nutrition of the immune system.

Nutrition of the Immune System

The maximal catalytic activities of a number of key enzymes in the metabolism of various fuels have been measured in rat and human lymphocytes and in mouse macrophages. These activities suggested that cells of the immune system are able to utilize, at high rates, glucose, glutamine and long-chain fatty acids. Rates of utilization of these fuels by isolated, incubated lymphocytes and by macrophages have been measured and they are indeed high. And of the glucose, glutamine and fatty acid utilized by these cells, very little is oxidized via acetyl-CoA and the classic Krebs cycle: glucose is converted almost totally into lactate, glutamine into glutamate, aspartate and lactate and fatty acids into ketone bodies and possibly other end-products [1, 2, 14]. Until this work, it had generally been considered that both lymphocytes and macrophages obtained most of their energy from glucose and, furthermore, that lymphocytes, which had not been subjected to an immune response (resting or quiescent lymphocytes), were metabolically inactive. A simple comparison shows the naivety of such a view: the rate of utilization of glucose plus glutamine by *resting* lymphocytes is about 25% of the rate of glucose utilized by the *maximally* physically working heart muscle (table 4).

This work has led to a new hypothesis to explain the high rates of partial oxidation of these fuels in lymphocytes and macrophages and, indeed, in neoplastic cells. It proposes that the high rates provide optimal conditions for *precise regulation* of the rates of purine and pyrimidine nucleotide synthesis, which utilize intermediates of both glycolysis and

Table 4. Rates of utilization of glucose or glutamine by lymphocytes and macrophages and other tissues in mouse, rat or man [data from 12, 14]

Animal	Tissue	Rates of utilization, nmol/h/mg protein	
		glucose	glutamine
Mouse	macrophage	355	186
Rat	mesenteric lymphocyte	42	223
	maximally-working heart	1,000	–
Man	peripheral lymphocyte	65	190
	brain	200	–

glutaminolysis when needed during proliferation of cells: the high rates provide a dynamic buffer for the use of the intermediates of these pathways for biosynthesis. This mechanism for providing precision in regulation is known as branched-point sensitivity in metabolic control [14]. Understanding the metabolic control theory that underlies this hypothesis is less important than appreciating the prediction to which it gives rise. It predicts that, if the plasma glutamine concentration is decreased below the physiological level, the rate of proliferation of lymphocytes and the function of macrophages will be decreased which would be expected to impair the function of the immune system. This prediction has been tested and shown to be true: thus lymphocytes in culture stimulated to proliferate by phytohaemagluttinin respond less well as the glutamine concentration is decreased [22].

The important point to emerge from this discussion is that glutamine must be used at a *high* rate for the cells of the immune system *even* when they are quiescent. The immune response to invasion by a micro-organism must be rapid: hence, the rate of glutamine utilization must always be high to provide optimal conditions for response to an immune challenge at *any* time. This then raises the question as to the source of this glutamine.

Muscle and the Immune System

Both liver and muscle can produce and release glutamine into the bloodstream. But muscle may be quantitatively the most important tissue. Since the cells of the intestine have a large capacity to utilize glutamine, most of the glutamine that enters the body via the diet is utilized by the intestine. Hence glutamine production and release by muscle becomes of considerable physiological and immunological importance.

At first site the situation appears to be somewhat more complex, since muscle appears to be able to take up glutamine and release it. Unfortunately, it appears to have been tacitly assumed that the *inward* transport of glutamine is the same process as that of glutamine *release* from muscle. Evidence has been obtained that uptake and release of glutamine by muscle are two separate processes [13]. This interpretation may have considerable physiological importance for regulation of the plasma glutamine level. In addition to this, skeletal muscle contains a high concentration of glutamine (20 mM in man) and this may have some important effects in muscle such as controlling the rate of protein synthesis. This potentially important effect of glutamine may also have directed attention away from the immunological significance of glutamine release by muscle.

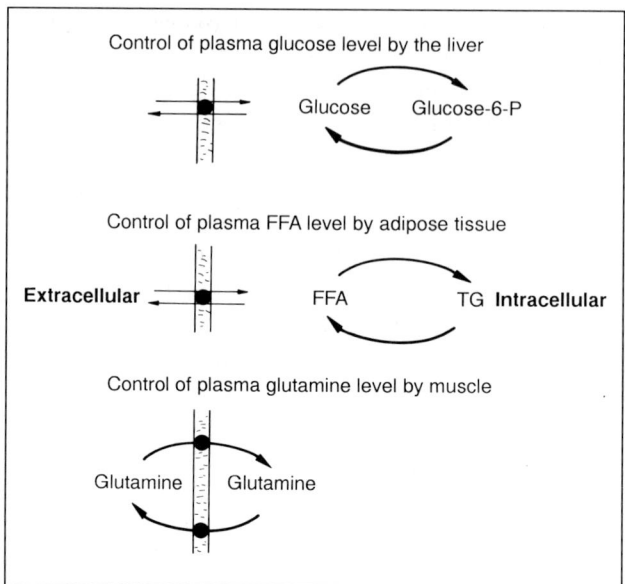

Fig. 1. Substrate/translocation cycles involved in the regulation of the plasma glucose, fatty acid and glutamine levels. The plasma glucose level is considered to be regulated, in part, by the substrate cycle between glucose and glucose-6-phosphate in the liver [12]; the plasma fatty acid level is considered to be regulated, in part, by the substrate cycle between fatty acid (fatty acyl-CoA) and triglyceride [12]; and the plasma glutamine level is considered to be regulated, in part, by the glutamine translocation cycle (see text).

If both glutamine uptake and glutamine release by muscle occur simultaneously in the same muscle fibre, then a translocation cycle exists for glutamine transport between intra- and extracellular glutamine pools (a translocation cycle is similar to a substrate cycle except that no chemical change occurs in glutamine). The advantage of this would be to confer enhanced sensitivity on the process of glutamine release from muscle and would provide a sensitive mechanism for the regulation of the blood glutamine level. This cycle in skeletal muscle would play a similar role to the glucose/glucose-6-phosphate cycle in liver for control of glucose release and the triglyceride/fatty acid cycle in control of the rate of release of fatty acid from adipose tissue (fig. 1).

Analysis of the processes of glutamine release by muscle and glutamine uptake by cells of the immune system by metabolic-control-logic

Table 5. Effect of different exercise regimens on plasma glutamine concentrations in man

Condition	Plasma glutamine concentration, μM					
	marathon race	30-km indoor race	30-km treadmill (run)	cycling		sprints (10×6 s)
				70% VO_{2max}	100% VO_{2max}	
Pre-exercise	592	532	641	558	510	556
Post-exercise	495**	503	694	581	502	616*

* $p < 0.05$ and ** $p < 0.005$, respectively, as compared with pre-exercise value (Student's t test). Data taken from Parry-Billings et al. [18, 19].

indicates that glutamine release from skeletal muscle is the flux-generating step in the pathway of glutamine utilization by cells of the immune system. Thus, the release process in muscle is considered to be nonequilibrium and to approach saturation with its substrate. Furthermore, the pathway of glutamine utilization in cells of the immune system is not saturated with this substrate so that changes in the rate of release of glutamine by muscle, via changes in the concentration of glutamine in plasma, will change the rate of glutamine utilization by immune cells. It follows that muscle glutamine release may, therefore, be the flux-generating step for the utilization of glutamine by the *immune system.* This emphasizes the important nutritional link between the two tissues. And, of importance, any insult to muscle may lead to an impairment of the translocation cycle involved in the control of glutamine release and hence in the rate of glutamine release by muscle. Could this be the explanation for the immunosuppression in overtrained state?

Effects of Exercise on Plasma Glutamine Concentrations

In human subjects the response of plasma glutamine to exercise varies according to the duration of the exercise. Short-term (6 s) sprints increase the plasma level of glutamine whereas endurance exercise (marathon race) decreases the level [18, 19] (table 5). Previous studies have reported similar results. Of interest, although the concentration of glutamine in plasma was decreased after a 42-km marathon, it was unaffected by exercise of 30-km duration. It is tempting to suggest that the difference between the two distances is the extent of glycogen depletion, and therefore, that deple-

Table 6. Effect of exercise on the contents of glutamine and the rates of glutamine and alanine release in isolated incubated soleus muscle of untrained and trained rats

Condition of animal	Content (µmol/g) of glutamine in soleus muscle after incubation		Rate of release from incubated (nmol/min/g) soleus muscle			
			glutamine		alanine	
	rested	post-exercise	rested	post-exercise	rested	post-exercise
Untrained	3.7	2.9	35.2	31.3	11.2	8.4**
Trained	1.4*	2.3**	37.2	33.0**	11.8	11.8

Data from Parry-Billings [17] are presented as means of at least nine separate experiments. Statistical significance: * $p < 0.05$ for difference between trained and untrained and ** $p < 0.05$ for difference between rested and exercised animals.

tion of muscle glycogen might be the cause of decreased rates of glutamine release. To some extent, similar results were observed in rats; plasma glutamine levels were decreased after exercise to exhaustion in trained rats, but were unchanged following exercise in sedentary rats, who ran for a *markedly* shorter time than the trained rats and may not have depleted their muscle glycogen: fatigue in these rats *may* have been caused by proton accumulation. When the soleus muscle was isolated from exhausted rats and incubated in vitro, the rate of glutamine release was decreased, in comparison to that of muscles removed from rested animals: the decrease in rate of release was, however, only statistically significant in the trained animals that ran much longer than the untrained animals (table 6) [17]. These findings support the view that the decrease in plasma glutamine level after endurance exercise is caused, at least in part, by a decreased rate of release from muscle: however, measurements of arteriovenous differences across skeletal muscle of rats and if possible man after exercise are essential to provide further support.

A Study of Athletes with the Overtraining Syndrome

The concentrations of glutamine, alanine and branched-chain amino acids have been measured in sedentary controls, trained athletes and overtrained athletes. The concentration of alanine and branched-chain amino acids were similar in the plasma of trained and overtrained athletes. However, of importance for the present discussion, the plasma concentration of

Table 7. Plasma amino acid concentrations in trained and overtrained subjects at rest[1]

		Plasma amino acid concentration, µM		
		glutamine	alanine	BCAA
Untrained		664	520	489
Trained		580	346	519
Overtrained		510*	350	490
Trained	'experimental'	608	498	357
	'follow-up'	621	542	472
Overtrained	'experimental'	503	438	333
	'follow-up'	535	487	394

* $p < 0.01$ as compared with the trained value (Student's t test). Data from Parry-Billings et al. [18, 19].
[1] Previously reported values from recreation of runners who were not highly trained have been included for comparison (and are described as 'untrained'). The lower section of the table shows the results for subsets of trained and overtrained subjects who were studied twice: firstly, when overtrained subjects presented with symptoms of overtraining ('experimental') and secondly, after a recovery period of 6 weeks ('follow-up'). The differences between experimental and follow-up values indicate that the recovery period had no significant effect on amino acid concentrations.

glutamine was *lower* in overtrained athletes compared with that in trained athletes and the concentration in trained subjects was lower than that in recreational runners (table 7) [18, 19].

Since samples were taken from resting subjects sometime after performance had been impaired, these results suggest that overtraining may have a long-term effect specifically on the plasma glutamine levels. Furthermore, following a 6-week recovery period, despite significant improvement in the exercise performance of these subjects, the plasma glutamine concentration remained below control values (table 7). This suggests that immunosuppression due to overtraining may persist for longer periods than indicated by the decrease in physical performance. Consequently, it is suggested that failure of muscle to provide enough glutamine could result in an impairment of the function of the immune system via lack of precision for the regulation of, for example, the rates of purine and pyrimidine nucleotide synthesis for DNA and RNA formation in lymphocytes. It will be important to investigate whether supplementation of the diet with glutamine or glutamine-peptides or adequate maintenance of glycogen levels in the muscle by adequate carbohydrate in the diet could prevent the immunosuppression of the overtrained state.

References

1 Ardawi MSM, Newsholme EA: Glutamine metabolism in lymphoyctes of the rat. Biochem J 1983;212:835–842.
2 Ardawi MSM, Newsholme EA: Metabolism in lymphocytes and its importance in the immune response. Essays Biochem 1985;21:1–44.
3 Berk LS, Nieman DC, Tan SA, Lee JW, Eby WC: Complement and immunoglobulin levels in athletes vs. sedentary controls. Proc Int Conf Exercise, Fitness, Health, Toronto 1988.
4 Blomstrand E, Celsing F, Newsholme EA: Changes in concentration of aromatic and branched chain amino acids during sustained exercise in man and their possible role in fatigue. Acta Physiol Scand 1988;133:115–121.
5 Blomstrand E, Perrett D, Parry-Billings M, Newsholme EA: Effect of sustained exercise on plasma amino acid concentrations and on 5-hydroxytryptamine metabolism in six different brain regions in the rat. Acta Physiol Scand 1989;136:473–481.
6 Fitzgerald L: Exercise and the immune system. Immunol Today 1988;9:337–339.
7 Keast D, Cameron K, Morton AR: Exercise and the immune response. Sports Med 1988;5:248–267.
8 Kuipers H, Keizer HA: Overtraining in elite athletes. Sports Med 1988;6:79–92.
9 Lightman SL, Everitt BJ: Neuroendocrinology. Oxford, Blackwell Scientific Publications, 1986.
10 Lewicki R, Tchorzewski H, Majewska E, Nowak Z, Baj Z: Effect of maximal physical exercise on T-lymphocyte sub-populations and on interleukin-1 (IL-1) and interleukin-2 (IL-2) production in vitro. Int J Sports Med 1988;9:114–117.
11 Nash HL: Can exercise suppress reproductive hormones in men? Physician Sports Med 1987;15:180–189.
12 Newsholme EA, Leech AR: Biochemistry for the Medical Sciences. Chichester, Wiley, 1983.
13 Newsholme EA, Parry-Billings M: Properties of glutamine release from muscle and its importance for the immune system. J Parent Enter Nutr 1991;14:635–675.
14 Newsholme EA, Newsholme P, Curi R, Challoner MA, Ardawi MSM: A role for muscle in the immune system and its importance in surgery, trauma, sepsis and burns. Nutrition 1988;4:261–268.
15 Noakes TD: The Lore of Running. Cape Town, Oxford University Press, 1986.
16 Oshida Y, Yamanouchi K, Hayamizu S, Sato Y: Effect of acute physical exercise on lymphocyte sub-population in trained and untrained subjects. Int. J Sports Med 1988;9:137–140.
17 Parry-Billings M: Studies on glutamine release by skeletal muscle; DPhil. Thesis Oxford University (1989).

18 Parry-Billings M, Blomstrand E, McAndrew N, Newsholme EA: A communicational link between skeletal muscle, brain and cells of the immune system. Int J Sports Med 1990;11:S122–S128.
19 Parry-Billings M, Blomstrand E, Leighton B, Dimitriadis GD, Newsholme EA: Does endurance exercise impair glutamine metabolism? Can J Sport Sci 1990;13:13P.
20 Rang HP, Dale MM: Pharmacology. Edinburgh, Churchill Livingstone, 1987.
21 Ryan AJ, Brown RL, Frederick EC, Falsetti HL, Burke ER: Overtraining in athletes. Physician Sports Med 1983;11:93–110.
22 Szondy Z, Newsholme EA: The effect of glutamine concentration on the activity of carbomyl-phosphate synthase II and on the incorporation of [^{3}H]thymidine into DNA in rat mesenteric lymphocytes stimulated by phytohaemagglutinin. Biochem J 1989;261:979–983.
23 Verma SK, Mahindroo SR, Kansal DK: Effect of four weeks of hard physical training on certain physiological and morphological parameters of basketball players. J Sports Med 1978;18:379–384.
24 Wagenmakers AJM, Brookes JH, Coakley JH, Reilly T, Edwards RHT: Exercise-induced activation of the branched-chain 2-oxo-acid dehydrogenase in human muscle. Eur J Appl Physiol 1989;59:159–167.

Dr. E.A. Newsholme, Department of Biochemistry, University of Oxford, South Parks Road, GB–Oxford OX1 3QU (UK)

Exercise, Plasma Composition, and Neurotransmission

Richard J. Wurtman[a], *Martin C. Lewis*[b]

[a] Department of Brain and Cognitive Sciences, and [b] Clinical Research Center, Massachusetts Institute of Technology, Cambridge, Mass., USA

Until a few decades ago, it was widely believed that the amount of its transmitter that any neuron released when it fired was more or less constant, so that if variations were to occur in the flow of information across the neuron's synapses, this would have to have resulted either from changes in the neuron's firing frequency or from processes involving receptors for its transmitter. But then it was shown that, for at least *some* of the 30-odd transmitters now known to exist, the *amount* of transmitter released each time the neuron fires also normally varies. Some of the processes underlying this variability were found to be mediated by inhibitory receptors located on the neuron's own presynaptic surface. Another important presynaptic process which modulates neurotransmission is the basis of this article, i.e., changes in neurotransmitter *synthesis* resulting from the metabolic consequences of normal activities, like eating or exercise. These activities produce their effects by raising or lowering plasma levels of nutrients which are the precursors of the neurotransmitter, in turn increasing or decreasing the amounts of transmitter released with each depolarization.

Nutrients now known to affect neurotransmission include choline, the precursor for acetylcholine; tryptophan and tyrosine, the precursors for serotonin and the catecholamines; and the other large neutral amino acids (LNAA), which compete with circulating tryptophan and tyrosine for transport into the brain.

Changes in neurotransmission caused by eating and by exercise can thus affect all of the behavioral and physiological functions that precursor-dependent neurons happen to subserve. The 'wrong' diet can theoretically *impair* exercise performance by making the athlete sleepy, or diminishing his ability to focus attention, or even by slowing the passage of nerve impulses to and along skeletal muscle. Conversely, it is possible that special nutrient mixtures can be designed to *enhance* particular types of neurotransmission just as though the nutrients were drugs. The biochemical processes which underlie these relationships are described below, as are some of their possible consequences for people engaged in athletics.

Dietary Carbohydrates and Proteins and Brain Serotonin

The intial observation that food consumption could affect neurotransmitter synthesis was made in studies on rats performed in 1971 [21]. Animals were allowed to eat a test diet which contained carbohydrates and fat but lacked protein. Soon after the start of the meal, brain levels of the essential (and scarce) amino acid tryptophan were found to have risen, thus increasing the substrate saturation of an enzyme, tryptophan hydroxylase, which controls serotonin synthesis. The resulting increase in brain serotonin levels was associated with an increase in brain levels of serotonin's chief metabolite, 5-hydroxyindole acetic acid (5-HIAA), suggesting that serotonin *release* had also been enhanced.

Initially difficult to explain were subsequent observations on what happened to brain tryptophan and serotonin levels after rats consumed a meal rich in protein: Although plasma tryptophan levels rose, reflecting the contribution of some of the tryptophan molecules in the protein, brain tryptophan and serotonin levels either failed to rise or, if the meal contained a high enough proportion of protein, actually fell [22]. The explanation for this paradox was found to lie in the properties of the transport molecules that carry tryptophan across the blood-brain barrier and into neurons [37]. The endothelial cells which line central nervous system capillaries are able to shuttle specific nutrients (or their metabolites) between the blood and the brain's extracellular space. One such transport molecule mediates the transcapillary flux (by facilitated diffusion) of tryptophan, tyrosine and other LNAA; others move choline, basic or acidic amino acids, hexoses, monocarboxylic acids, adenosine, adenine, and various vitamins. The transport molecule which ferries the LNAA into and out of the brain attaches its ligands competitively; each LNAA competes with the others. (Transport of the portion of plasma tryptophan (about 80%) which circulates loosely bound to albumin occurs almost as efficiently as that of 'free' tryptophan (because the affinity for the amino acids is much greater than the affinity of albumin); hence, in most circumstances the ratio of plasma 'total' tryptophan to other LNAA best predicts brain tryptophan level [30].)

Hence the ability of circulating tryptophan molecules to enter the brain is increased when plasma levels of the other LNAA fall (as occurs after insulin is secreted) [33] and diminished when the other LNAA rise, even if plasma tryptophan levels remain constant. Since all dietary proteins are considerably richer in the other LNAA than in tryptophan (which represents only 1.0–1.5% of most proteins), and since some of the LNAA, unlike tryptophan, are only poorly metabolized within the portal system, consumption of a protein-rich meal decreases the 'plasma tryptophan ratio' (the ratio of the plasma tryptophan concentration to the summed concentrations of its major circulating competitors for brain uptake: tyrosine, phenylalanine; the branched-chain amino acids leucine, isoleucine, and valine; methionine). This, in turn, decreases tryptophan's transport into the brain and slows its conversion to serotonin. (Similar competitive mechanisms also mediate the fluxes of tryptophan and other LNAA between the brain's extracellular space and its neurons; moreover, similar plasma ratios also can be used to predict brain levels of tyrosine and the other LNAA after meals or other treatments – like exercise – which modify plasma amino acid patterns [12].

It seems counterintuitive that the meal which most effectively raises brain tryptophan levels should be the one that lacks tryptophan entirely (that is, one containing carbohydrates but no proteins), while a protein-rich meal, which elevates blood tryptophan concentrations substantially, has the opposite effect on the brain. But this relationship has been documented again and again. Plasma tryptophan ratios in normal individuals vary between about 0.065 and 0.160, depending largely on the composition of the last meal (or snack) eaten, and the interval that has passed since its ingestion [23, 31]. In rats, such variations are capable of causing sizable differences in brain tryptophan levels. Subnormal plasma tryptophan ratios are often noted in obese people, reflecting elevated plasma levels of the branched-chain amino acids, caused in part by insulin resistance and in part by the increase in *lean* body mass also characteristic of obese people [10, 11, 25]; these ratios are further reduced if the subjects are put on a high-protein, low-carbohydrate diet. Apparently, no data are available on plasma tryptophan ratios and LNAA concentrations in weight lifters, who have increased lean body mass but lack obesity.

The fact that giving pure tryptophan can increase brain serotonin synthesis, and can thereby affect various serotonin-dependent brain functions (sleepiness, mood) had been known for a decade [42]. What was novel and perhaps surprising about the above experiments was their demonstration that brain tryptophan levels – and serotonin synthesis – *normally* undergo important variations, for example, in response to the decision to eat a carbohydrate-rich versus a protein-rich breakfast (or perhaps, to undertake heavy exercise).

It remained possible, however, that mechanisms might exist beyond the serotonin-releasing neuron itself, which might keep precursor-induced increases in transmitter synthesis from causing parallel changes in the amounts of serotonin actually released into synapses. Indeed, it was known that if rats were given *very large* doses of tryptophan – sufficient to raise brain tryptophan levels well beyond their normal range – the firing frequencies of their serotonin-releasing raphe neurons decreased markedly; this was interpreted as reflecting the operation of a feedback system designed to keep serotonin release within a physiologic range. However, if rats were given *small* doses of tryptophan – sufficient to raise brain tryptophan levels, but not beyond their *normal* peaks – or if they consumed a carbohydrate-rich meal, which raised brain tryptophan levels *physiologically,* no decreases in raphe firing occurred. Hence, food-induced changes in serotonin synthesis *are* able to affect the amounts of serotonin released per firing without slowing the neuron's firing frequencies, and foods probably are 'allowed' to modulate the net output of information from serotoninergic neurons. (This output is, theoretically, the product of three factors: the number of serotonin-releasing nerve terminals; the average frequency with which the raphe neurons happen to be firing; and the average number of serotonin molecules released at each terminal per firing.) The physiologic range of serotonin release per unit time apparently is very broad.

Tryptophan, 'Nutritional Supplements', Dietary Carbohydrates, and the Human Brain

The ability of supplemental tryptophan to enhance serotonin turnover within the human's central nervous system (for example, to elevate CSF 5-HIAA levels) was first shown in 1970 [19]; apparently, no neurochemical data are available on the human brain's responses to carbohydrate intake. Numerous behavioral and neurologic effects have been shown to be associated with tryptophan administration, starting with Smith and Prockop's [42] original observation that it caused drowsiness. Most of these effects have been reviewed extensively elsewhere and are not further discussed here.

Pure tryptophan has never been developed in the United States as a legitimate drug; rather, the amino acid has been marketed in health-food stores [41] as a 'nutritional supplement'. Its sale for that purpose can only be regarded as cynical: The *only* clinical circumstance in which a person might be shown to have an isolated tryptophan deficiency would be the very rare patient with a widely metastasized carcinoid tumor. Giving pure tryptophan to someone with a general protein deficiency would only make the consequence of that deficiency worse. Not labelling supplemental tryptophan as what it really is, i.e., an over-the-counter drug, has meant that its production has not been governed by 'Good Manufacturing Practices' standards, and that consumers have not been given package inserts listing the amino acid's contraindications and side effects, its approved uses ('indications'), and the doses to be used. In many ways, this method of distributing tryptophan was akin to an accident waiting to happen. The accident did in fact happen in 1989, when many cases of a sometimes fatal 'eosinophilia-myalgia' syndrome were identified among people taking the amino acid purchased from health-food stores [41]. At the time of writing it remains uncertain whether this man-made disease results from impurities in the tryptophan, or from the use of inappropriately high dosages, or perhaps from individual metabolic idiosyncrasies among the afflicted patients. We submit that *no* nutrient which can be shown to have physiological or behavioral effects should be allowed to be marketed as a 'nutritional supplement'. However, many such compounds still are: Numerous additional accidents are waiting to happen ... Only a few well-controlled studies have been published describing behavioral effects of dietary carbohydrates [43]. Some of these have involved administering a specific carbohydrate, sugar (sucrose), to hyperactive children whose parents or teachers believed that this carbohydrate exacerbated their behavioral problem. In general, consumption of the sugar tended, if anything, to *reduce* activity, similar to its (and tryptophan's) reported effect on normal individuals: A high-carbohydrate lunch increased sleepiness in women, calmness in men, and, in subjects over 40, the tendency to commit errors in a standardized test of performance.

Brain Serotonin, Nutrient Choice, and Carbohydrate Craving

If a rat is allowed to pick from foods in two pans, presented concurrently, containing differing proportions of protein and carbohydrate, it chooses between the two so as to obtain fairly constant (for each animal) amounts of the macronutrients, However, if prior to 'dinner' it receives either a carbohydrate-based 'snack' [50] or a drug that facilitates serotoninergic neurotransmission [51], it quickly modifies its food choice, selectively diminishing its intake of carbohydrates.

These observations support the hypothesis that the ability of serotoninergic neurons to modulate their release of transmitter in reponse to food-induced changes in plasma amino acids allows these neurons to serve a special function, as 'sensors', in the brain's mechanisms governing nutrient choice [47]: Probably they participate in a feedback loop through which the composition of 'breakfast' (that is, its proportions of protein and carbohydrate) can – by increasing or decreasing brain serotonin levels – influence the choice of 'lunch'.

A similar mechanism seems to operate in humans: Subjects housed in a research hospital were allowed to choose from six different isocaloric foods (containing varying proportions of protein and carbohydrate, but constant amounts of fat) at each meal, taking as many small portions as they liked; they also had continuous access to a computer-driven vending machine, stocked with mixed carbohydrate-rich and protein-rich isocaloric snacks. It was observed that the basic parameters of each person's food intake – total number of calories; grams of carbohydrate and protein; number and composition of snacks – tended to vary within a narrow range, day to day, and to be unaffected by placebo administration.

To assess the involvement of brain serotonin in maintaining this constancy of nutrient intake, pharmacologic studies were undertaken in individuals in whom this putative regulatory mechanism was thought to be impaired. These were obese people who claimed to suffer from 'carbohydrate craving', manifested as their tendency to consume large quantities of carbohydrate-rich snacks, usually at a characteristic time of day or evening [53].

Subjects were given dexfenfluramine, a drug which had been found to decrease carbohydrate intake in normal rats, and to cause weight loss in obese people by a mechanism involving the release of brain serotonin [51]. Administration of relatively low doses (15 mg twice daily) caused a major reduction in snack carbohydrate intake; a smaller reduction in mealtime carbohydrates; and no significant changes at all in mealtime protein or fat intake [52, 53]. (Too few protein-rich snacks were consumed by the subjects to allow assessment of the drug's effect on this source of calories).

Other drugs also thought to enhance serotonin-mediated neurotransmission selectively, for example, the antidepressants zymelidine, fluvoxamine and fluoxetine, also have been found to cause weight loss; this contrasts with the weight gain (and carbohydrate craving) often associated with less chemically specific antidepressants like amitriptyline. It has not yet been determined whether these drugs also selectively suppress carbohydrate intake in humans. Severe carbohydrate craving is also characteristic of patients suffering from SADS, a variant of bipolar clinical depression associated with an onset in Fall; a higher frequency in populations living far from the equator; and concurrent hypersomnia and weight gain [39]. Moreover, a reciprocal tendency of many obese people to suffer from mood disorders (usually depression) has also been noted [39]. Since serotoninergic neurons apparently are involved in the actions of both appetite-reducing and antidepressant drugs, they might constitute the link between a patient's appetitive and mood symptoms: Some patients with disturbed serotoninergic neurotransmission might consult their physicians for obesity, reflecting their overuse of dietary carbohydrate to treat their dysphoria. (The carbohydrates, by increasing intrasynaptic serotonin, would mimic the neurochemical actions of bona fide antidepressant drugs like the MAO inhibitors and tricyclic compounds). Other patients might bring to their physician a complaint of depression, and their carbohydrate craving and weight gain would be perceived as secondary problems. Another group of patients – the nonanorexic bulimics – might seek medical assistance because of their concurrent food binges and depression or anxiety. Yet other groups might include women with the premenstrual syndrome [8] exhibiting mood disturbances, weight gain, carbohydrate craving, and sometimes fluid retention; and people attempting to discontinue smoking [Spring et al., in press]. The participation of serotoninergic neurons in a large number of brain functions besides nutrient-choice regulation might have the effect of making such functions hostages to eating (seen in the sleepiness that can, for example, follow carbohydrate intake), just as it could cause mood-disturbed individuals to consume large amounts of carbohydrate for reasons related neither to the nutritional value nor taste of these foods. The incidence of carbohydrate craving in athletes – especially women – awaits study. The possible beneficial effects of carbohydrate on muscle glycogen stores might be counterbalanced by their tendency to make people less vigilant and even sleepy.

Choline and Lecithin: Effects on Acetylcholine Synthesis

The amounts of acetylcholine released by physiologically active cholinergic neurons depend on the concentrations of choline available to them. In the absence of supplemental free choline, the neurons will continue to release fairly constant quantities of the transmitter; however, when choline is made available (in concentrations bracketing the physiologic range), a clear dose relationship is observed between its concentration and acetylcholine release [7, 32]. (The biochemical mechanism that couples a cholinergic neuron's firing to its choline responsiveness awaits discovery.) When adequate free choline is unavailable, the source of the choline used for acetylcholine synthesis is the cells' own membranes [7, 32, 46]. Membranes are very rich in phosphatidylcholine (PC), and this phospholipid serves as a 'reservoir' of free choline, much as bone and albumin serve as reservoirs for calcium and essential amino acids.

Neurons can draw on three sources of free choline for acetylcholine synthesis: that stored as PC in their own membranes; that formed intrasynaptically from the hydrolysis of acetylcholine (and taken back up into the presynaptic terminal by a high-affinity process estimated to be 30-50% efficient in the brain and somewhat more so at the neuromuscular junction); and that entering the bloodstream from the diet or the liver and kidneys (and taken into the brain by a specific blood-brain barrier transport system). PC in foods (liver, eggs) is rapidly hydrolyzed to free choline in the intestinal mucosa (or broken down more slowly, after passage into the lymphatic circulation). Consumption of adequate quantities of PC can lead to severalfold elevations in plasma choline levels, thereby increasing brain choline and the substrate saturation of the enzymes that make acetylcholine and the PC in neuronal membranes [7].

Tyrosine Effects on Dopamine and Norepinephrine Synthesis

Catecholaminergic neurons apparently *become* tyrosine-sensitive when they are *physiologically active,* and *lose* this capacity when they are quiescent. The biochemical mechanism that couples a neuron's firing frequency to its ability to respond to supplemental tyrosine involves phosphorylation of the tyrosine hydroxylase enzyme protein, a process that occurs when the neurons fire. This phosphorylation, which is short-lived, enhances the enzyme's affinity for its cofactor (tetrahydrobiopterin) and makes it insensitive to end-product inhibition (by norepinephrine and other catechols): these changes allow its net activity to depend on the extent to which it is saturated with tyrosine.

In recent studies, we have used the in vivo microdialysis technique [17] to assess effects of tyrosine on the release of dopamine from the rat's corpus striatum. When otherwise untreated animals receive the amino acid systemically, there is, after 20–40 min, a substantial increase in dopamine output, unaccompanied by detectable increases in levels of dopamine's metabolites DOPAC or HVA [17]. But this effect is short-lived, dopamine release returning to basal levels after 20–30 min. This latter response is caused by receptor-mediated decreases in the firing frequencies of the striatal neurons (to compensate for the increase in dopamine release which occurs per firing), and by local presynaptic inhibition resulting from the transient increase in intrasynaptic dopamine levels. If animals are given haloperidol, a dopamine receptor-blocking agent, prior to or along with the tyrosine, the supplemental tyrosine continues to amplify dopamine output for prolonged periods [18].

The tight coupling of tyrosine responsiveness to neuronal firing frequency probably explains tyrosine's paradoxical effects on blood pressure: The amino acid *elevates* blood pressure (and sympathoadrenal catecholamine release) in hypotensive animals [12] but *lowers* blood pressure (without affecting sympathoadrenal catecholamines) in hypertensive animals [44]. (It fails to have any effect on blood pressure in normotensive animals or humans [34].) Tyrosine's blood pressure-lowering effect in hypertensive animals probably results from its conversion to norepinephrine in brain stem neurons, which, when active, suppress sympathetic outflow; these neurons presumably are activated in the types of hypertension in which tyrosine is effective, and may be involved in the brain's attempts to deal with the hypertension. As might be anticipated tyrosine administration elevates brain levels of the norepinephrine metabolite MHPG-SO_4 in these animals, but has little or no effect on brain MHPG-SO_4 in rats with normal or low blood pressure.

Supplemental tyrosine may also have value in the prophylaxis or treatment of stress responses, including those to the stress of severe exercise: Rats subjected to a standard laboratory stress were found, immediately thereafter, to have depressed brain norepinephrine levels (particularly in the locus coeruleus and hypothalamus), probably reflecting the inability of norepinephrine's synthesis to keep up with its release; they also showed behavioral abnormalities and elevated plasma corticosterone levels [38].

All of these changes, including the adrenocortical hypersecretion, were suppressed by supplemental oral tyrosine, but not if the tyrosine was coadministered with another LNAA (valine) that blocked its brain uptake.

Effects of Sustained Exercise on Plasma Nutrients Affecting Neurotransmitter Synthesis

The possibility that sustained exercise might affect the amounts of nutrients that would be available for conversion to neurotransmitter molecules was suggested in early studies [3, 20] showing increases in plasma tyrosine levels among subjects exercising vigorously on a cycle. This observation was not confirmed by other investigators [9, 26, 45] nor in subjects completing a 100-km race [15]. However, in the latter study the *ratio* of plasma tyrosine to other LNAA increased by about half because of the concurrent major decreases in plasma levels of leucine, isoleucine, and valine. Increases in the plasma tyrosine/LNAA ratio – as well as in tryptophan and the tryptophan/LNAA ratio – were also described in rats forced to run on a treadmill; these increases were associated with indirect evidence that the synthesis of brain serotonin also changes in parallel [1]. More recent studies suggest that sustained exercise can increase the levels of serotonin and its metabolite 5-HIAA within some regions of rat brain – and also those of the catecholamine neurotransmitters dopamine and norepinephrine (see below) [6].

To determine whether really major exercise might cause robust changes in the levels of plasma nutrients affecting neurotransmitters, we and our associates measured these nutrients in blood samples taken from highly-trained subjects before and immediately after they ran the Boston Marathon. In one experiment, involving 17 male subjects running the 1985 Marathon, plasma choline was measured; similar data were obtained on additional subjects running the 1986 Marathon. Blood samples from 37 male and female subjects running either the 1985 or 1986 Marathons were also assayed for tyrosine, tryptophan, and the other LNAAs (phenylalanine, leucine, isoleucine, valine, methionine).

Marathon running was found to be associated with a major decline (about 40%) in mean plasma choline, from 10.1 ± 0.5 to 6.2 ± 0.3 μM, a level well below that usually observed in fasting subjects [7], and probably sufficient [5] to cause a significant decline in the release of acetylcholine at the neuromuscular junction unless supplemental choline could be recruited from another source (e.g., the 'autocannibalism' of neuronal or muscular membrane phospholipids [46]. Similar observations were made in runners the following year, i.e., a fall in plasma choline from 14.1 ± 1.90 to $8.4 \pm 0.57\,M$. Marathon running thus induced what was apparently the first physiological condition found to *decrease* plasma choline levels; the magnitude of this decrease was comparable to the fall in plasma choline subsequently observed at Boston University [55] in 11 normal subjects given a virtually choline-free diet for 3 weeks (i.e., from 12.2 to 7.4 μM, or about 40%). (It is interesting to note that the diet-induced reduction in plasma choline was associated not with an impairment in neuromuscular transmission, but with a slowing in the transmission of contraction-generating impulse across the skeletal muscle itself, an effect compatible with altered membrane phospholipid metabolism.) No studies have been done, to date, to determine whether supplementation with a compound that releases choline into the blood-stream – like a choline salt, or PC – can block the exercise-induced decline in plasma choline, and perhaps thereby improve either exercise performance or acetylcholine-dependent brain functions. Studies are now in progress to examine these possibilities. The biochemical mechanism responsible for the fall in plasma choline also awaits identification. It could involve, among other possibilities, the conversion of 'free choline' (i.e., the moiety which was measured) to high-density lipoproteins, or perhaps a decrease in hepatic blood flow, with a consequent decrease in the amount of choline secreted by this organ in the fasting state.

Marathon running also was found to cause significant *increases* in plasma levels of tyrosine, phenylalanine, and methionine – all of which are principally metabolized in the liver – and in their corresponding ratios to the other LNAA [14], while *decreasing* the plasma ratios of leucine, isoleucine and valine – amino acids metabolized not in liver but in skeletal muscle. Greatest increases were observed in tyrosine (47% and in the tyrosine/LNAA ratio (40%). Plasma tryptophan levels and the tryptophan/LNAA ratio increased slightly but not significantly in these studies. Conceivably, the exercise-induced rise in tyrosine availability might have enhanced the ability of the sympathoadrenal system [12], and the locus coeruleus [27] to continue to make and release their catecholamine transmitters in spite of the stress of the Marathon. Supplemental tyrosine (100 mg/kg) has, in fact, been shown to enhance mental performance, improve the mood, and diminish symptoms in human subjects exposed to such stressors as cold and high altitude [4]. It also enabled subjects to withstand lower body negative pressure for longer periods, and produced electroencephalographic changes (in the amplitude of the P300 wave) consistent with increased decision-making ability [16]. Tyrosine's utility in enhancing exercise performance is currently being examined in a joint research program involving the Massachusetts Institute of Technology, the Institute of Cardiology of the Soviet Union, and the Soviet Olympic Riflery team.

When supplemental choline is given to a long-distance runner, the intention is to correct a deficiency state caused by the exercise, and thereby restore acetylcholine-mediated neurotransmission to normal. In contrast, when supplemental tyrosine is given to an athlete, the intention is not to restore a deficiency but to obtain a pharmacological effect, i.e., the release of additional amounts of norepinephrine from the locus coeruleus of the brain, and perhaps elsewhere. Both kinds of goals may be satisfied in a new generation of special food-based products for athletes.

Conclusion

The levels of amino acids and of choline in the plasma can control the rates at which the monoamine neurotransmitters and acetylcholine are synthesized in, and released from, nerve cells in the brain and at the neuromuscular junction. Thus, changes in plasma composition caused, for example, by eating are able to modulate transmitter release, with carbohydrate consumption enhancing serotonin release, and tyrosine or choline ingestion increasing that of norepinephrine or of acetylcholine.

Since strenuous exercise also is able to affect plasma amino acid and choline levels, it is likely that exercise also modulates neurotransmission, perhaps secondarily affecting performance. Prolonged heavy exercise (e.g., marathon running, bicycling, cross-country skiing) lowers plasma levels of the branched-chain amino acids and raises those of tyrosine and tryptophan. In rats, such changes in plasma composition can be shown to increase serotonin release in the brain.

Marathon running also decreases plasma choline levels and may thereby affect both central nervous system and skeletal muscle function: choline decreased from 10.1 ± 0.5 to $6.2 \pm 0.3\,M$ (mean $\pm$ SEM) among 17 runners in the 1985 Boston Marathon [13], and a similar decrease (from 14.1 ± 1.90 to $8.4 \pm 0.57\,M$) was observed among runners in the 1986

Marathon [unpubl. observations]. The potential functional consequences of this effect are under investigation.

The consumption, prior to sustained exercise, of particular macronutrients chosen to affect plasma amino acid or choline levels may provide a way to enhance performance significantly. These compounds could act either as nutrients, meeting the special nutritional requirements generated by heavy exercise, or as drugs, raising their own plasma levels beyond the normal range and thus promoting maximal neurotransmitter release. Choline is an example of the former, and tyrosine – now being tested for vigilance-enhancing properties – of the latter.

References

1. Acworth I, Nicholass J, Morgan B, Newsholme E: Effect of sustained exercise on concentrations of plasma aromatic and branched-chain amino acids and brain amines. Biochem Biophys Res Commun 1986;137:149–153.
2. Agharanya J, Alonso R, Wurtman RJ: Changes in catecholamine excretion following tyrosine ingestion in normally fed human subjects. Am J Clin Nutr 1981;34:82–87.
3. Ahlborg G, Felig P, Hagenfeldt L, Handler R, Wahren J: Substrate turnover during prolonged exercise in man. J Clin Invest 1974;53:1080–1090.
4. Banderet LE, Lieberman HR: Treatment with tyrosine, a neurotransmitter precursor, reduced environmental stress in humans. Brain Res Bull 1989;22:759–762.
5. Bierkamper GG, Goldberg AM: in Wurtman JJ, Wurtman RJ (eds): Nutrition and the Brain. New York, Raven Press, 1979, vol 5, pp 243–251.
6. Blomstrand E, Perrett D, Parry-Billings M, Newsholme EA: Effect of sustained exercise on plasma amino acid concentrations and on 5-hydroxytryptamine metabolism in six different regions in the rat. Acta Physiol Scand 1989;136:473–481.
7. Blusztajn JK, Wurtman RJ: Choline and cholinergic neurons. Science 1983;221:614–620.
8. Blusztajn JK, Lopez G, Coviella I, Logue M, Growdon JH, Wurtman RJ: Levels of phospholipid catabolic intermediates, glycerophosphocholine and glycerophosphoethanolamine are elevated in brains of Alzheimer's disease but not of Down's syndrome patients. Brain Res, in press.
9. Brodan V, Kuhn E, Pechar J, Tomkova J: Changes of free amino acids in plasma of healthy subjects induced by physical exercise. Eur J Appl Physiol 1976;35:69–77.
10. Caballero B, Finer N, Wurtman RJ: Plasma amino acids and insulin levels in obesity: Response to carbohydrate intakes and tryptophan supplements. Metabolism 1988;37:672–776.
11. Caballero B, Wurtman RJ: Differential effects of insulin resistance on leucine and glucose kinetics in obesity. Metabolism, in press.
12. Conlay LA, Maher TJ, Wurtman RJ: Tyrosine increases blood pressure in hemorrhagic shock. Science 1981;212:559–560.

13 Conlay L, Wurtman RJ, Blusztajn JK, Lopez G, Coviella I, Maher TJ, Evoniuk GE: Marathon running decreases plasma choline concentrations in neonatal plasma. N Engl J Med 1986;315:892.
14 Conlay L, Wurtman RJ, Maher TJ, Lopez GCI, Blusztajn JK, Vacanti CA, Logue M, During M, Caballero B, Evoniuk G: Effects of running the Boston Marathon on plasma concentrations of large neutral amino acids in human subjects. J Neural Transm 1989;76:65–71.
15 DeCombaz J, Reinhardt P, Anantharaman K, von Glutz G, Poortmans JR: Biochemical changes in a 100-km run: free amino acids, urea, and creatinine. Eur J Appl Physiol 1979;41:61–72.
16 Dollins AB, Krock LP, Storm WF, Lieberman HR: Tyrosine decreases physiological stress caused by lower body negative pressure (abstract). Aviat Space Environ Med 1990;6:491.
17 During M, Acworth IN, Wurtman RJ: Effects of systemic tyrosine on dopamine release from rat striatum and nucleus accumbens. Brain Res 1989;452:378–380.
18 During M, Acworth IN, Wurtman RJ: Dopamine release in rat striatum: Physiological coupling to tyrosine supply. J Neurochem 1989;52:1449–1454.
19 Eccleston D, Ashcroft GW, Crawford TBB, Stanton JB, Wood D, McTurk PH: Effect of tryptophan administration on 5-HIAA in cerebrospinal fluid in man. J Neurol Neurosurg Psychiatry 1970:33:269.
20 Felig P, Wahren J: Amino acid metabolism in exercising man. J Clin Invest 1971;50: 2703–2714.
21 Fernstrom JD, Wurtman RJ: Brain serotonin content: physiological dependence on plasma tryptophan levels. Science 1971;173:149–152.
22 Fernstrom JD, Wurtman RJ: Brain serotonin content: physiological regulation by plasma neutral amino acids. Science 1972;178:414–416.
23 Fernstrom JD, Wurtman RJ, Hammarstrom-Wiklund B, Rand WM, Munro HN, Davidson CS: Diurnal variations in plasma concentrations of tryptophan, tyrosine, and other neutral amino acids: effect of dietary protein intake. Am J Clin Nutr 1979; 32:1912–1922.
24 Gibson CJ, Wurtman RJ: Physiological control of brain norepinephrine synthesis by brain tyrosine concentration. Life Sci 1978;22:1399–1406.
25 Heraief E, Burckhardt P, Mauron C, Wurtman J, Wurtman RJ: The treatment of obesity by carbohydrate deprivation suppresses plasma tryptophan and its ratio to other large amino acids. J Neural Transm 1983;57:187–195.
26 Holm G, Bjorntorp P, Jagenburg R: Carbohydrate, lipid, and amino acid metabolism following physical exercise in man. J Appl Physiol 1978;45:128–131.
27 Lehnert H, Reinstein DK, Strowbridge BW, Wurtman RJ: Neurochemical and behavioral consequences of acute, uncontrollable stress: Effects of dietary tyrosine. Brain Res 1984;303:215–223.
28 Little A, Levy R, Chaqui-Kidd P, Hand D: A double-blind placebo-controlled trial of high-dose lecithin in Alzheimer's disease. J Neurol Neurosurg Psychiatry 1985; 48:736–742.
29 Lovenberg W, Ames MM, Lerner P: Mechanism of short-term regulation of tyrosine hydroxylase; in Lipton MA, DiMascio A, Killam KF (eds): Psychopharmacology: A Generation of Progress. New York, Raven Press, 1978, pp 247–259.
30 Madras BK, Cohen EL, Messing R, Munro HN, Wurtman RJ: Relevance of serum-

free tryptophan to tissue tryptophan concentrations. Metabolism 1974;23:1107–1116.
31 Maher TJ, Glaeser BS, Wurtman RJ: Diurnal variations in plasma concentrations of aspartate and glutamate: effects of dietary protein intake. Am J Clin Nutr 1984;39: 722–729.
32 Maire JC, Wurtman RJ: Effects of electrical stimulation and choline availability on release and contents of acetylcholine and choline in superfused slices from rat striatum. J Physiol (Paris) 1985;80:189–195.
33 Martin-Du-Pan R, Mauron C, Glaeser B, Wurtman RJ: Effect of increasing oral glucose doses on plasma neutral amino acid levels. Metabolism 1982;31:937–943.
34 Melamed E, Glaeser B, Growdon JH, Wurtman RJ: Plasma tyrosine in normal humans: effects of oral tyrosine and protein-containing meals. J Neural Transm 1980;47:299–306.
35 Milner JD, Wurtman RJ: Catecholamine synthesis: physiological coupling to precursor supply. Biochem Pharmacol 1986;35:875–881.
36 Nasrallah HA, Dunner FL, Smith RE, McCalley-Whitters M, Sherman AD: Variable clinical response to choline in tardive dyskinesia. Psychol Med 1984;14:697–700.
37 Pardridge WM: Regulation of amino acid availability to the brain; in Wurtman RJ, Wurtman JJ (eds): Nutrition and the Brain. New York, Raven Press, 1977, vol 7, pp 141–204.
38 Reinstein DK, Lehnert H, Wurtman RJ: Dietary tyrosine suppresses the rise in plasma corticosterone following acute stress in rats. Life Sci 1985;37:2157–2163.
39 Rosenthal NE, Heffernan MM: Bulimia, carbohydrate craving, and depression: a central connection; in Wurtman RJ, Wurtman JJ (eds): Nutrition and the Brain. New York, Raven Press, 1985, vol 7, pp 139–166.
40 Scally MC, Ulus I, Wurtman RJ: Brain tyrosine level controls striatal dopamine synthesis in haloperidol-treated rats. J Neural Transm 1977;41:1–6.
41 Slutsker L, Hoesley SC, Miller L, Williams LP, Watson JC, Fleming DW: Eosinophilia-myalgia syndrome associated with exposure to tryptophan from a single manufacturer. JAMA 1990;264:213–217.
42 Smith B, Prockop DJ: Central nervous system effects of ingestion of L-tryptophan by normal subjects. N Engl J Med 1962;267:1338–1341.
43 Spring B: Effects of foods and nutrients on the behavior of normal individuals; in Wurtman RJ, Wurtman JJ (eds): Nutrition and the Brain. New York, Raven Press, 1985, vol 7, pp 1–47.
44 Sved AF, Fernstrom JD, Wurtman RJ: Tyrosine administration reduces blood pressure and enhances brain norepinephrine release in spontaneously-hypertensive rats. Proc Natl Acad Sci USA 1979;76:3511–3514.
45 Tysper Z, Kosicki B, Waliszewski K: Plasma free amino acid levels in subjects during maximal workload. Acta Physiol Pol 1980;31:101–106.
46 Ulus I, Wurtman RJ, Mauron C, Blusztajn JK: Choline increases acetylcholine release and protects against the stimulation-induced decrease in phosphatide levels within membranes of rat corpus striatum. Brain Res 1989;484:217–227.
47 Wurtman RJ: Behavioral effects of nutrients. Lancet 1983;i:145–147.
48 Wurtman RJ: Alzheimer's disease. Sci Am 1985;252:62–75.
49 Wurtman RJ, Larin F, Mostafapour S, Fernstrom JD: Brain catechol synthesis: control of brain tyrosine concentration. Science 1974;185:183–184.

50　Wurtman JJ, Moses PL, Wurtman RJ: Prior carbohydrate consumption affects the amount of carbohydrate that rats choose to eat. J Nutr 1983;113:70–78.
51　Wurtman JJ, Wurtman RJ: Drugs that enhance central serotoninergic transmission diminish elective carbohydrate consumption by rats. Life Sci 1979;24:895–904.
52　Wurtman JJ, Wurtman RJ, Growdon JH, Lipscomb HP, Zeisel SA: Carbohydrate craving in obese people: suppression by treatments affecting serotoninergic transmission. Int J Eating Disord 1981;1:2–15.
53　Wurtman JJ, Wurtman RJ, Mark S, Tsay R, Wilbert W, Growdon J: D-Fenfluramine selectively suppresses carbohydrate snacking by obese subjects, Int J Eating Disord 1985;4:89–99.
54　Yuwiler A, Oldendorf WH, Geller E, Braun L: Effect of albumin binding and amino acid concentration on tryptophan uptake into brain. J Neurochem 1977;28:1015–1023.
55　Xia, Nan: Effects of dietary choline levels on human muscle function; MS thesis Boston University College of Engineering, pp 1–71.

Richard J. Wurtman, MD, Department of Brain and Cognitive Sciences, Massachusetts Institute of Technology, Cambridge, MA 02139 (USA)

L-Carnitine Supplementation and Performance in Man

Anton J.M. Wagenmakers

Department of Human Biology, Nutrition Research Centre,
University of Limburg, Maastricht, The Netherlands

Introduction

The use of supplementary L-carnitine by athletes has become widespread in recent years, especially since anecdotes have been circulating that this substance has helped the Italian soccer team to become world champion in 1982 [1]. However, not many unequivocal scientific data have been published indicating that athletes are at risk of developing L-carnitine deficiency and have an increased need for L-carnitine. It also has not been shown unequivocally that L-carnitine supplementation improves performance in healthy human subjects. In this review critical reflections will be made on a number of controversial experimental data that have appeared in the scientific literature. We also will reflect on the biochemical mechanisms which have been proposed for a potential ergogenic effect of this compound. We only will cover those aspects of L-carnitine metabolism which are of relevance to athletic performance in man and we only will refer to those studies which are properly controlled and not biassed by mismatches in experimental groups and inproper endpoints in performance tests. For a more complete review on the biochemistry of L-carnitine and its role in physiology and pathophysiology, we refer to earlier reviews [2, 3].

Dietary Intake, Biosynthesis, Body Stores and Losses

Man obtains L-carnitine from the diet (particularly red meats and dairy products) and endogenous biosynthesis. The relative contributions of endogenously synthesized and dietary L-carnitine have not been established to date, but it is generally assumed that, when the diet does not contain L-carnitine, enough L-carnitine appears to be produced in healthy individuals to maintain homeostasis. For this reason, L-carnitine is not regarded as a vitamin, but as a vitamin-like substance. Dietary L-carnitine is absorbed by the small intestine both by active transport and passive diffusion [4]. L-Carnitine is synthesized in mammalian cells from trimethyllysine. Specific lysinyl residues (in intracellular proteins) are methylated in the 6-amino position. This reaction also requires the essential amino acid methionine (methylation via S-adenosyl-methionine). After protein degradation the resulting trimethyllysine is released and metabolized to L-carnitine via four enzymatic steps. The first three steps are currently thought to occur in the tissue of origin and lead to the production of trimethylammoniobutanoate [5]. In humans the final reaction occurs in liver, kidney and brain but not in heart and skeletal muscle [5]. The latter tissues contain high concentrations of L-carnitine and therefore are dependent upon transport through the circulation for L-carnitine. Muscle and many other cell types take L-carnitine up against a concentration gradient (plasma total L-carnitine concentrations about 40–60 μM; muscle concentrations 3–4 mM) by a saturable active transport process [6]. In a 70-kg man the total body L-carnitine store is close to 100 mmol. Ninety-eight percent of this is in skeletal and heart muscle, 0.6% in extracellular fluid and 1.6% in liver and kidney [6]. An intracellular pathway of L-carnitine degradation has not been identified in man and higher mammals. L-Carnitine is lost from the body via urine and stool. The daily (urinary) excretion of L-carnitine is 100–300 μmol [2], which is 0.1–0.3% of the total body store. Meat-free diets reduce urinary losses to below 100 μmol/day. Fecal losses usually account for less than 1% of urinary losses [6].

In tissues and physiological fluids, L-carnitine is present in the free form and in esterified forms as short-chain and long-chain acyl-L-carnitines. The predominant short-chain acyl-L-carnitine ester is the water-soluble acetyl-L-carnitine. Other short-chain acyl-L-carnitines identified in human blood and urinie are propionyl-, butyryl-, isobutyryl- and isovaleryl-L-carnitine [7]. In man at rest, acyl-L-carnitine esters account for 10–40% of total L-carnitine in plasma and for 4–33% of total L-carnitine in muscle and liver. Acyl-L-carnitine esters may constitute as much as 50–60% of total L-carnitine in urine. These proportions may vary considerably with nutritional status [8] and as indicated below before and after exercise.

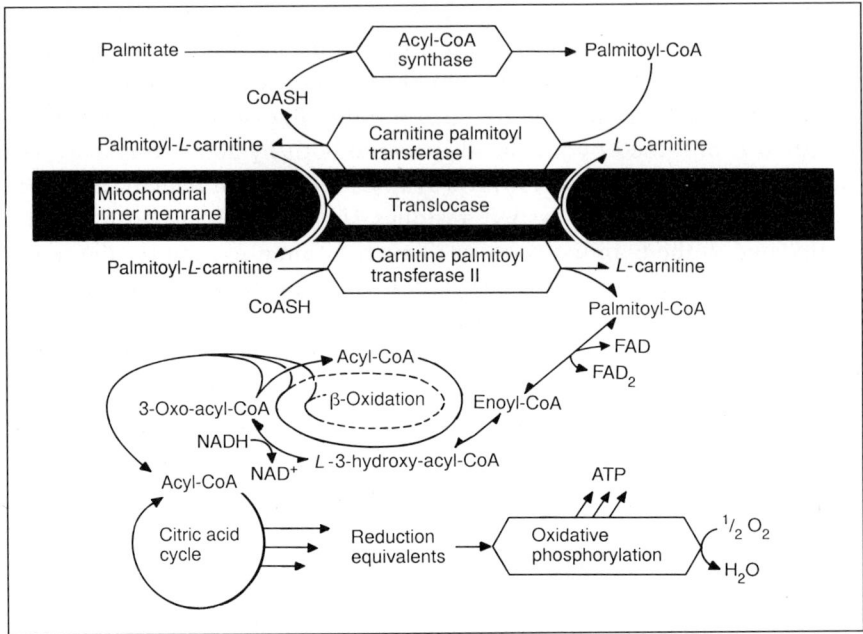

Fig. 1. Carnitine-mediated transport of long-chain acyl-CoA across the inner mitochondrial membrane [from 43].

Metabolic Role of L-*Carnitine and Theoretical Considerations for Oral Supplementation in Athletes*

Enhancement of Long-Chain Fatty Acid Oxidation

The primary function of *L*-carnitine is to transfer long-chain fatty acids across the inner mitochondrial membrane [9] (fig. 1). Cytosolic long-chain fatty acids are activated to their coenzyme A (CoA) esters by fatty acyl-CoA synthetase and subsequently transesterified to *L*-carnitine by carnitine palmitoyltransferase I, an enzyme located on the external surface of the inner mitochondrial membrane [10]. Transport of the long-chain fatty acyl-*L*-carnitine esters across the inner mitochondrial membrane is mediated by acylcarnitine translocase and occurs in exchange for free *L*-carnitine of mitochondrial origin [11]. On the matrix side of the inner mitochondrial membrane long-chain acyl-CoA is then regenerated by the action of carnitine palmitoyltransferase II and the produced fatty acyl-CoA subsequently undergoes β-oxidation. Due to this essential role in long-chain fatty acid oxidation, L-carnitine supplementation has been suggested as a means to enhance fatty acid oxidation during prolonged exercise.

There are some considerations why an enhanced fatty acid oxidation could be of benefit for athletes. During exercise both carbohydrates and fatty acids are used to meet the energy requirements of the athlete. However, as the glycogen stores in the muscle become depleted the rate of glycolysis and hence that of glucose oxidation will decrease progressively. It has been argued that the maximal rate at which fatty acids can be used during aerobic exercise is limited to about 50% of the maximum aerobic rate of ATP production [12, 13]. This means that when the glycogen stores are depleted in the runner or cyclist, the pace must fall by 50%. In addition an increased contribution of fatty acid oxidation to energy expenditure could reduce the rate of glycogen breakdown and thus postpone fatigue [12, 13]. Addition of L-carnitine to the incubation medium can markedly increase long-chain fatty acid oxidation by isolated muscle mitochondria or muscle homogenates [9]. If oral supplementation of L-carnitine would have the same effect in vivo during exercise, then it could increase the contribution of fatty acid oxidation to total energy expenditure, spare glycogen breakdown and postpone fatigue. Furthermore, a higher maximal rate of fatty acid oxidation would be possible and could increase the pace that can be maintained by the athlete after glycogen stores have been emptied. However, L-carnitine only can have these effects in vivo when there is a relative shortage of L-carnitine and when the reactions for which L-carnitine is a substrate are limiting the maximal rate of fatty acid oxidation. The K_m of carnitine palmitoyltransferase for L-carnitine is 250–450 μm [2] which implicates that the muscle L-carnitine concentration under normal circumstances should be high enough to allow the enzyme to function at maximal activity. Furthermore, under normal circumstances the exchange capacity of carnitine acylcarnitine translocase most likely exceeds the capacity of mitochondria to oxidize fatty acids with a wide margin [2]. So, the reactions in long-chain fatty acid oxidation for which L-carnitine is a substrate do not seem to be rate-limiting unless athletes would be at risk of developing a shortage of muscle free L-carnitine during exercise (see below). The rate of supply of fatty acids (i.e., by the concentration of free fatty acids in the circulation) rather has been suggested as the limiting factor for the rate of ATP formation from fatty acid oxidation [2, 12–14] during prolonged exercise. Furthermore, we recently proposed an alternative mechanism for the development of fatigue [15, 16], which may further explain why L-carnitine in vivo would not limit fatty acid oxidation. Exercise in general leads to an acceleration of the breakdown of the branched-chain amino acids (BCAA; leucine, isoleucine and valine). The first step in the degradation pathway of the BCAA is an aminotransferase reaction, in which the citric acid (Krebs) cycle intermediate 2-oxoglutarate is used as amino group acceptor (fig. 2). Activation of BCAA metabolism may, therefore, drain the citric acid cycle and limit oxidation of fuels. The draining effect of the BCAA aminotransferase reaction normally will be counteracted by the carbon flux from glycogen and pyruvate into the citric acid cycle (fig. 2). This flux, however, is limited in endurance exercise leading to glycogen depletion. According to this mechanism the flux in the citric acid cycle will thus fall in prolonged exercise and limit the oxidation of fatty acids at a step beyond the L-carnitine-dependent transport into the mitochondria.

Modulation of the Acetyl-CoA/CoA Ratio and the Size of the CoA Pool

Acetyl-CoA can be converted to acetyl-L-carnitine in the following reaction:

Acetyl-CoA + L-carnitine ↔ acetyl-L-carnitine + CoA

The reaction is freely reversible, with an equilibrium constant equal to 0.6–0.7, and is catalyzed by the enzyme carnitine acetyltransferase [2]. It has been suggested that oral L-carnitine supplementation, by increasing the muscle L-carnitine concentration, could liberate free CoA and decrease the acetyl-CoA/CoA ratio and thus stimulate the activity of pyruvate dehydrogenase during exercise and enhance the in vivo utilization of glucose [17–19]. There is little doubt that this mechanism is involved in the L-carnitine-induced stimulation of pyruvate oxidation observed in vitro in incubated mitochondria [17], but the relevance of this mechanism to the in vivo exercise situation is not clear. A decrease in the acetyl-CoA/CoA ratio would interfere with the inhibitory action of increases in fatty acid oxidation on glycogen and glucose oxidation. This action is a very potent mechanism to spare glycogen and thus postpone fatigue, once fatty acids have been mobilized and are available for oxidation in muscle. It is mediated by increases in acetyl-CoA as an end-product of β-oxidation [12]. L-Carnitine supplementation could thus theoretically enhance glycogen breakdown and induce premature fatigue, when it indeed would lead to a decrease of the acetyl-CoA/CoA ratio in prolonged endurance exercise. However, a

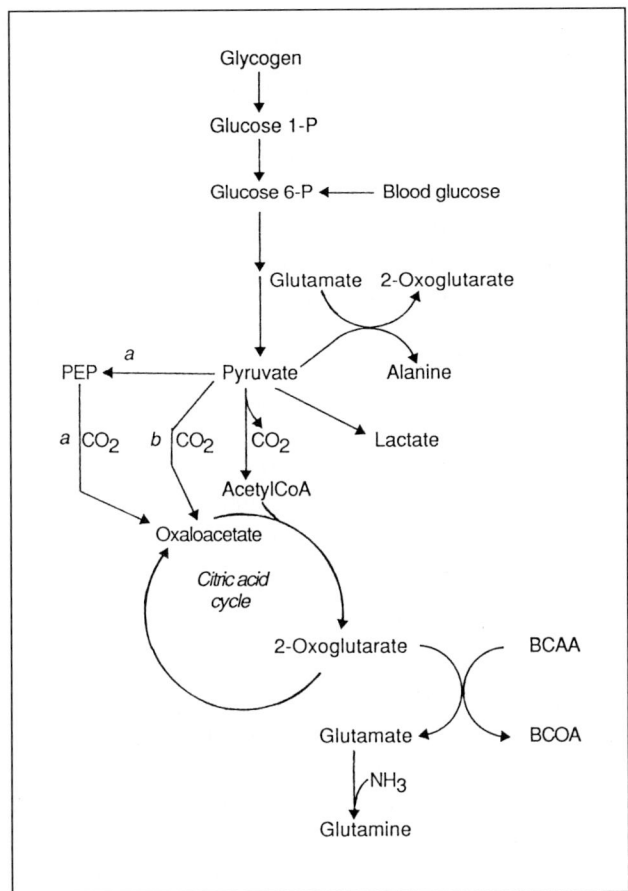

Fig. 2. Alternative biochemical mechanism of muscle fatigue. Exercise in general leads to an acceleration of the breakdown of the BCAA (leucine, isoleucine and valine). The first step in the degradation pathway of the BCAA is an aminotransferase reaction, in which the citric acid cycle intermediate 2-oxoglutarate is used as amino group acceptor. Activation of BCAA metabolism may, therefore, drain the citric acid cycle and limit oxidation of fuels. The draining effect of the BCAA aminotransferase reaction is normally counteracted by the anaplerotic conversion of glycogen and glucose to citric acid cycle intermediates via carboxylation of pyruvate (route *b*) and/or reversal of the phosphoenolpyruvate carboxykinase reaction (route *a*). This flux, however, is limited in endurance exercise leading to glycogen depletion. According to this mechanism the flux in the citric acid cycle will thus fall in prolonged exercise and limit the oxidation of fatty acids. BCOA = branched-chain 2-oxo acids; PEP = phosphoenolpyruvate [from 16].

decrease of the acetyl-CoA/CoA ratio could be of relevance for short maximal and supramaximal exercise such as the 100-m sprint. In that case, lactate is the normal end-product of anaerobic glycolysis and accumulation of lactate and the parallel decrease of the pH play an important role in the development of fatigue [12] and also reduce the maximal rate of glycolysis by inhibition of phosphofructokinase [12]. If pyruvate oxidation would be stimulated by an L-carnitine-induced decrease of the acetyl-CoA/CoA ratio then less lactate would be formed and more acetyl-L-carnitine during short intense exercise and performance could improve in that way.

The potential increase of free CoA as a consequence of an increase in muscle L-carnitine also has been suggested to improve performance during prolonged exercise by making more free CoA available for the 2-oxoglutarate → succinyl-CoA step in the citric acid cycle, thus enhancing the metabolic flux through this cycle. Again there is little doubt that this mechanism is involved in the L-carnitine-induced stimulation of the oxidation of 2-oxoglutarate to succinate observed in vitro [20], but its in vivo relevance is not clear.

Enhancement of BCAA Oxidation

Finally, it has been suggested that L-carnitine could stimulate the oxidation of the BCAA during exercise [18]. Again convincing evidence has been presented that this indeed is the case in vitro [21, 22], probably by conversion of branched-chain acyl-CoA esters to branched-chain acyl-L-carnitines thus removing the feedback inhibition of the branched-chain acyl-CoA esters on the branched-chain 2-oxo acid dehydrogenase complex and making free CoA available for other metabolic purposes. If this mechanism would apply to the in vivo situation then L-carnitine supplementation could have a negative effect on performance since it could lead to an increased breakdown rate of the BCAA, an increased drain of the BCAA aminotransferase reaction on 2-oxoglutarate, a reduction of the flux in the citric acid cycle and premature fatigue (fig. 2) [15, 16].

In conclusion, several rationales have been given in the past for potential ergogenic effects of oral L-carnitine supplementation. Critical reflections indicate that some of these mechanisms have not been worked out properly and on theoretical grounds may turn out to have no or even a negative effect instead of being ergogenic. Furthermore, all the proposed mechanisms assume that oral L-carnitine supplementation will lead to increases of the muscle L-carnitine concentration in man. To the best of our knowledge this has not been investigated (see below).

Effects of Exercise on Muscle, Plasma and Urinary L-Carnitine and Its Esters

Muscle

Muscle contains about 98% of the body total L-carnitine stores. Estimates of potential exercise-induced changes in the body stores therefore should include muscle concentrations. Lennon et al. [23] reported that total muscle L-carnitine fell by about 20% in 28 subjects when cycling for 40 min at 55% $\dot{V}_{O_2max}$. Carlin et al. [24] studied 10 male subjects during 90 min of cycling exercise at 50% $\dot{V}_{O_2max}$ and did not find a change in total muscle L-carnitine after exercise. Free L-carnitine fell from 2.90 to 1.85 µmol/g wet weight and esterified L-carnitine increased from 0.80 to 1.79 µmol/g wet weight. Hiatt et al. [25] studied 6 male subjects both at low intensity exercise (60 min at about 40% $\dot{V}_{O_2max}$) and high intensity exercise (30 min at 87–89% $\dot{V}_{O_2max}$). With low intensity exercise there was a (nonsignificant) 16% decrease of total muscle L-carnitine, while the short-chain acyl-L-carnitine fraction remained unchanged. Ten minutes of high intensity exercise decreased total muscle L-carnitine significantly with 19%. No further change occurred between 10 and 30 min of high intensity exercise. Muscle short-chain acyl-L-carnitine content increased fivefold, contributing 9% of total L-carnitine at rest and 60–67% between 10 and 30 min of high intensity exercise. Free L-carnitine showed a compensatory fall. After 60 min of recovery, total muscle L-carnitine had returned to baseline values, but short-chain acyl-L-carnitine content remained elevated and free L-carnitine remained decreased. Janssen et al. [26] measured total muscle L-carnitine before and after a marathon (n = 20) and found a (nonsignificant) 7% fall. They [26] also showed that resting total muscle L-carnitine did not change during a 1-year training period of volunteers, ending with a marathon. It is known that intense dynamic exercise leads to 10–20% increases of the total water content of muscle [27, 28] diluting all the solutes in muscle, including total L-carnitine. The observed decreases of total muscle L-carnitine, therefore, most likely do not represent a loss of L-carnitine or L-carnitine esters from the muscle during exercise (a claim made only by Lennon et al. [23]). This potential dilution problem would not exist when concentrations are expressed on dry weight basis. In agreement with this, Harris et al. [29] did not observe changes in total muscle L-carnitine content expressed per kilogram dry weight in 3 subjects who cycled for 4 min at 90% $\dot{V}_{O_2max}$. However, free L-carnitine fell from about 77% of total L-carnitine at rest to about 37%

after exercise and acetyl-*L*-carnitine increased from 19% at rest to about 50% after exercise. In 1 of the subjects, acetyl-*L*-carnitine increased to 99% of total *L*-carnitine. Soop et al. [30] studied the exchange of fuels, *L*-carnitine and its esters during 120 min of exercise at 50% $\dot{V}_{O_2max}$ in 7 subjects by measuring the femoral arteriovenous differences and leg blood flows. The leg released a minute amount of free *L*-carnitine (2.5 μmol/min) after 120 min of exercise, but no release occurred after 40 min of exercise. There was no leg release of acyl-*L*-carnitines after 40 and 120 min of exercise.

In conclusion, these studies indicate that there is no substantial loss of *L*-carnitine and its esters from the exercising muscles. During low and moderate intensity exercise there appear to be no or relatively small changes in the proportion of acylated and free *L*-carnitine in muscle. During high intensity exercise the free *L*-carnitine concentration falls from 77–90% at rest to 30–37% [25, 29]. This fall is compensated for by a proportional increase in short-chain acyl-*L*-carnitines and primarily due to an increase of acetyl-*L*-carnitine [29]. These figures indicate that acetyl-*L*-carnitine in addition to lactate is a major metabolite formed during high intensity exercise and suggest that *L*-carnitine may function in the regulation of the acetyl-CoA/CoA ratio by buffering excess production of acetyl-CoA [29].

Plasma

No differences have been observed in the resting plasma concentration of total *L*-carnitine between sedentary individuals and endurance trained athletes [19, 31, 32]. Total plasma *L*-carnitine levels do not change substantially [23–25, 30, 31, 33] after exercise of various duration and intensity. The free *L*-carnitine fraction decreases in most of these studies and is counterbalanced or even exceeded by an increase of the acyl-*L*-carnitine fraction. According to Soop et al. [30], who studied the leg exchange of *L*-carnitine and its esters during exercise, the acyl-*L*-carnitines are released at a site other than the contracting muscles.

Urine

Urinary losses of *L*-carnitine increased from 340 to 580 μmol/day in 3 Japanese students after severe running exercise [34]. Exercise of various intensities and duration did not change the urinary *L*-carnitine excretion when normalized for creatinine excretion [25, 30]. Since extracellular fluids contain only 0.6% of the body *L*-carnitine store and since daily urinary losses normally account for less than 0.3% of the body L-carnitine store, much bigger exercise-induced changes would have to occur in these variables in order to be indicative of an increased L-carnitine requirement of athletes.

Effect of L-*Carnitine Supplementation on Plasma, Muscle and Urinary* L-*Carnitine and Its Esters*

Several authors [30, 31, 33, 35] have investigated the effect of oral supplementation of L-carnitine (2–5 g/day for periods of 5–14 days). Plasma resting total L-carnitine concentration increased in three of these studies with a value of 36–75% [30, 31, 35]. No change was observed by Cooper et al. [33] in marathon runners. Siliprandi et al. [36] gave 2 g of L-carnitine orally and found a 32% increase in plasma total L-carnitine after 1 h. Similar increases were observed in all these studies [30, 31, 35, 36] for the resting plasma levels of free L-carnitine and acylated L-carnitine with no change in the partition between these fractions. Effects of oral L-carnitine supplementation on muscle L-carnitine concentration to the best of our knowledge have not been investigated in man. Lennon et al. [37], however, investigated the relationship between dietary L-carnitine intake and plasma and muscle L-carnitine concentrations in 28 volunteers. There was a significant positive relationship between dietary intake and plasma total L-carnitine concentrations, but no relationship between dietary intake and muscle total L-carnitine concentrations. This may indicate that increased plasma concentrations not necessarily lead to an increased muscle concentration. This also would be predicted on theoretical grounds, since, as indicated before, muscle and many other cell types take L-carnitine up against a concentration gradient (plasma total L-carnitine concentrations about 40–60 μM; muscle concentrations 3–4 mM) by a saturable active transport process [6]. Soop et al. [30] studied the femoral arteriovenous exchange of L-carnitine after 5 days of oral supplementation of 5 g L-carnitine/day and did not find uptake by the muscle at rest and after 120 min of exercise at 50% $\dot{V}_{O_2max}$ despite a 75% increase in the plasma concentration of total L-carnitine; after 40 min of exercise there was a significant release of L-carnitine. The urinary excretion of L-carnitine (per 24 h) increased about tenfold after L-carnitine supplementation [30, 36]. In normal subjects the renal plasma excretory threshold for both total and free L-carnitine is close to the normal plasma L-carnitine concentration (about $50\,\mu M$) [6]. This implicates that increases of plasma L-carnitine observed after oral supplementation automatically will lead to an increase of the urinary excretion. To the best of our knowledge no L-carnitine balance studies have been performed during oral L-carnitine supplementation, so that it is not known how much of the oral supplement is retained in the body and how much is lost via urine and stool.

In conclusion, no direct measurements have been made of muscle L-carnitine concentrations after oral supplementation, but dietary intake of L-carnitine did not correlate with the muscle concentration and oral supplementation of L-carnitine did not lead to uptake by the leg muscles.

Effect of L-*Carnitine Supplementation on Aerobic and Anaerobic Performance*

Marconi et al. [31] studied 6 competitive endurance runners during aerobic and anaerobic exercise. Oral supplementation of L-carnitine (4 g/day for 2 weeks) increased the $\dot{V}_{O_2max}$ significantly by 6%. No effects were found on $\dot{V}_{O_2}$ and respiratory exchange ratio during 2 h of exercise at 65% $\dot{V}_{O_2max}$. Increases in plasma lactate concentration during supramaximal exercise were similar after L-carnitine supplementation. The authors concluded that the $\dot{V}_{O_2max}$ was increased as a result of activation of the substrate flow through the citric acid (Krebs) cycle, whereas the contribution of fatty acid oxidation to metabolism in prolonged submaximal exercise remained unchanged. Greig et al. [35] did not find an effect of oral L-carnitine (2 g/day for 2–4 weeks) on $\dot{V}_{O_2max}$ in two trials with 9 and 10 fit untrained individuals. Values of cardiac frequency at 50% of $\dot{V}_{O_2max}$ were significantly reduced in trial 1 but not in trial 2. Cardiac frequency at 75% and 100% $\dot{V}_{O_2max}$ showed no significant differences in either trial. Postexercise plasma lactate concentration was not changed by the L-carnitine supplement. Soop et al. [30] studied 7 moderately trained human subjects during 2 h of cycling exercise at 50% $\dot{V}_{O_2max}$. Fuel oxidation was estimated by measuring among others the femoral arteriovenous exchange of fatty acids, glucose, lactate and leg blood flow. L-Carnitine supplementation (5 g/day for 5 days) did not influence fatty acid and glucose oxidation, lactate production or utilization of other substrates either at rest or during prolonged exercise. Heart rate during exercise after L-carnitine supplementation decreased significantly with about 7–8% ($p < 0.05$), but O_2 uptake was unchanged. Leg blood flow increased after L-carnitine with

8.4%, but this difference was not significant. Siliprandi et al. [36] studied 10 moderately trained subjects in a double-blind cross-over design during incremental exercise. L-Carnitine (2 g) or placebo were given 1 h before the test. L-Carnitine increased total work done, decreased postexercise plasma lactate and pyruvate concentrations with about 10% and increased postexercise plasma acetyl-L-carnitine concentration from about 10 to 20 μM. Unfortunately the exercise sessions were stopped in this study when subjects achieved the theoretical maximal heart rate (220 − age). This is not a proper endpoint especially not if L-carnitine would affect the heart rate.

Discussion and Conclusions

In conclusion, oral L-carnitine supplementation does not appear to increase the rate of fatty acid oxidation during prolonged exercise. This observation is in agreement with the earlier reasoning that the maximal rate of fatty acid oxidation probably would be limited by L-carnitine-independent steps. The changes in muscle free L-carnitine during prolonged exercise indeed were small [23, 24, 30], so that enough L-carnitine appears to be available in human muscle to allow maximal activities of carnitine palmitoyltransferase and carnitine acylcarnitine translocase. Only during brief intense exercise free L-carnitine fell substantially due to formation of acetyl-L-carnitine [25, 29], but aerobic oxidation of fatty acids does not contribute significantly to energy provision in that situation. In fact, the accumulation of acetyl-CoA and acetyl-L-carnitine indicate that the flux in the citric acid cycle is limiting aerobic oxidation of fuels in that situation, so that there is no need to accelerate the L-carnitine-dependent transport of fatty acids into the mitochondria.

The substantial production of acetyl-L-carnitine during brief intense exercise, on the other hand, does indicate that L-carnitine may help to modulate the acetyl-CoA/CoA ratio in that situation. A lower acetyl-CoA/CoA ratio may increase the activity of pyruvate dehydrogenase, thus reducing accumulation of pyruvate and lactate and potentially improving performance in that way. A claim that L-carnitine supplementation may improve performance during incremental exhaustive exercise by way of this mechanism has been made by Siliprandi et al. [36] on the basis of decreases in postexercise plasma lactate and pyruvate concentrations and a 10 μM increase in postexercise plasma acetyl-L-carnitine (note that the increase in muscle acetyl-L-carnitine concentration during brief intense exercise is in the mM range in unsupplemented individuals [25, 29]). L-Carnitine supplementation, however, did not lead to lower blood lactate values after supramaximal, maximal and submaximal exercise in other studies [30, 31, 35] and also did not change the leg release of lactate and acetyl-L-carnitine during submaximal exercise [30]. To answer the question whether L-carnitine supplementation may enhance performance by stimulation of pyruvate dehydrogenase activity during maximal and supramaximal exercise, we in fact would need data on lactate release and muscle biopsy concentrations of lactate, L-carnitine and acetyl-L-carnitine in those situations, but to the best of our knowledge these data are not available.

In fact we also urgently need data on the effect of L-carnitine supplementation on muscle L-carnitine levels before scientific conclusions can be reached on the potential mechanism of the very few (controversial) positive effects which have been reported. If L-carnitine supplementation does not increase the muscle-free L-carnitine concentration then it does not make sense to speculate on intracellular biochemical mechanisms. What about the 6% increase of $\dot{V}_{O_2max}$ reported by Marconi et al. [31] in highly trained individuals, which could not be reproduced by others in healthy untrained subjects [35]. What about the significant 3–4% decrease in heart rate observed by Greig et al. [35] at 50% $\dot{V}_{O_2max}$ in trial 1, but not in trial 2? What about the 7–8% decrease in heart rate observed by Soop et al. [30] at 50% $\dot{V}_{O_2max}$? Are these real effects, which are too small to be caught consistently in laboratorium trials? If they are, then, due to missing scientific data, we cannot decide at the moment whether they are due to intramuscular mechanisms or extracellular pharmacological effects.

In this respect it is interesting to mention recent results obtained by Dubelaar et al. [38] in an in situ muscle fatigue test in the dog. L-Carnitine was given by infusion and increased serum L-carnitine from 23 to 322 µM. L-Carnitine administration increased the maximal force developed upon electrical stimulation with 31% and increased the total amount of work performed by the muscle. The effect was specific for L-carnitine (no effect was observed with D-carnitine and the structural analogue choline). The muscle L-carnitine concentration was measured and did not change upon L-carnitine administration. The authors, therefore, suspected an extracellular mode of action of L-carnitine and therefore also measured blood flow. L-Carnitine increased blood flow both at rest and during exercise [38]. Peripheral capillary dilation, enhancing oxygen supply and substrate and metabolite exchange, may well explain the data of Dubelaar et al. [38], the increase in $\dot{V}_{O_2max}$ observed by Marconi et al. [31] and the decrease in heart rate observed by Greig et al. [35] and Soop et al. [30] at 50% $\dot{V}_{O_2max}$. Future research will be required to investigate this potential mechanism and provide the definitive answer whether oral supplementation of L-carnitine can be of use to athletes by this or alternative mechanisms.

Toxic Effects of D-Carnitine and D,L-Carnitine Mixtures

Besides the biologically active L-carnitine isomer, synthesized in mammals, a second enantiomeric D-isomer does exist. The first chemical methods for the synthesis of carnitine gave rise to synthesis of the enantiomeric mixture D,L-carnitine. The first preparations available to athletes in fact were D,L-carnitine mixtures. D-carnitine, however, causes depletion of L-carnitine in mammalian tissues and may inhibit carnitine acyltransferase enzymes [39], thus leading to toxic effects in man. Myasthenia-like syndromes were reported in dialysis patients treated with D,L-carnitine for 20 days [40]. Use of oral D,L-carnitine (two times 500 mg daily for 2 days before a half marathon) caused myoglobinuria and extreme and continued weakness of the quadriceps muscles in a trained runner after the race [41], symptoms similar to those described for carnitine deficiency patients. So, it is clear that the use of D,L-carnitine is dangerous for athletes and for those using oral carnitine supplements it, therefore, is advised to check carefully the purity of the preparations that they use. Since then, purer and ultrapure L-carnitine has become available on an industrial scale by separation of the D- and L-isomer and by new synthetic routes giving rise solely to L-carnitine [42]. Despite this improvement in the chemical synthesis of L-carnitine on an industrial scale, commercial preparations of L-carnitine these days may still contain substantial D-carnitine impurities. It is advised not to use L-carnitine preparations without specification of the producer (quality control may have been poor when source is not specified) and of the % D-carnitine impurity and to use only preparations with an enantiomeric purity of $\geq 99.0\%$. The percentage of D- and L-carnitine can be estimated by measurement of the specific optical rotation and by ^{1}H-NMR [42].

Supplementation with Lysine and Methionine Does Not Increase L-Carnitine Biosynthesis

As indicated before, L-carnitine is synthesized in mammalian tissues from trimethyllysine. Net trimethyllysine synthesis involves the two essential amino acids lysine and methionine. For this reason, supplements of these amino acids have been used in man in order to increase the endogenous synthesis of L-carnitine, but this does not make sense. The free amino acid lysine cannot be used for synthesis of trimethyllysine but only specific lysinyl residues in intracellular proteins. These are methylated in a reaction requiring the presence of methionine (via S-adenosyl-methionine). Under normal circumstances the human diet contains enough methionine, so that it also does not make sense to increase the methionine intake of athletes. In general, supplementation of the diet with individual amino acids should be avoided, since it may cause an imbalance in the dietary supply of amino acids and could lead to adverse effects.

Practical Advice to Athletes and Medical Advisers on Oral Supplementation of L-*Carnitine*

We do not feel that at the moment there is a scientific basis for the use of L-carnitine by athletes. It has unambiguously been shown that L-carnitine supplementation does not increase fatty acid oxidation during prolonged exercise, which was the original rationale for its use. It may have ergogenic effects in other ways, but in this respect there are more questions than answers and a lot of research has to be done to provide definitive answers and potential mechanisms. However, this kind of scientific advice may have little influence on medical sport advisers, coaches and athletes and therefore on the use of L-carnitine in sport practice. It therefore seems sensible to consider the following details.

L-Carnitine is a water-soluble vitamin-like product. Megadoses of the ($\geq 99.0\%$ enantiomeric) pure product, therefore, should be relatively harmless and if no retention in the body occurs then the excess will be excreted in the urine. Despite this it is not known whether the continuous use of megadoses is without risk to man. Therefore, if one decides nevertheless to use the compound, it should only be used intermittently during a period of days up to weeks. Daily doses of 2–5 g L-carnitine have been used in scientific experiments [30, 31, 33, 35] for 5 days up to 4 weeks without adverse effects. Larger amounts either on single occasions or continuously should never be taken until it has been shown scientifically that this is harmless and without risks. Professional cyclists who are competing daily for several months should at least stop L-carnitine supplementation intermittently for several days and to stop at least in the winter season.

References

1 Editorial: Münch Med Wochenschr 1982;124:9.
2 Bremer J: Carnitine – Metabolism and functions. Physiol Rev 1983;63:1421–1466.
3 Borum PB: Clinical Aspects of Human Carnitine Deficiency. Oxford, Pergamon Press, 1986.
4 Hamilton JW, Li BUK, Shug AL, Olsen WA: Carnitine transport in human intestinal biopsy specimens. Gastroenterology 1986;91:10–16.
5 Hoppel CL, Davis AT: Inter-tissue relationship in the synthesis and distribution of carnitine. Biochem Soc Trans 1986;14:673–674.
6 Engel AG, Rebouche CJ: Carnitine metabolism and inborn errors. J Inherited Metab Dis 1984;7(suppl 1):38–43.
7 Valkner KJ, Bieber LL: Short-chain acylcarnitines of human blood and urine. Biochem Med 1982;28:197–202.
8 Bieber LL, Emaus R, Valkner K, Farrell S: Possible functions of short-chain and medium-chain carnitine acyltransferases. Fed Proc 1982;41:2858–2862.
9 Fritz IB: The metabolic consequences of the effects of carnitine on long-chain fatty acid oxidation; in Gran FC (ed): Cellular Compartmentalization and Control of Fatty Acid Metabolism. New York, Academic Press, 1968, pp 39–63.
10 Hoppel CL: Carnitine and carnitine palmitoyltransferase in fatty acid oxidation and ketosis. Fed Proc 1982;41:2853–2857.
11 Pande SV, Parvin R: Characterization of carnitine acylcarnitine translocase system of heart mitochondria. J Biol Chem 1976;251:6683–6691.
12 Newsholme EA, Leech AR: Biochemistry for the Medical Sciences. Chichester, Wiley, 1983.
13 Newsholme EA: Application of metabolic logic to the questions of causes of fatigue in marathon races; in Macleod D, Maughan R, Nimmo M, Reilly T, Williams C (eds): Exercise: Benefits, Limits and Adaptations. London, E & FN Spon, 1987, pp 181–198.
14 Carlson LA, Pernow B: Studies on blood lipids during exercise. I. Arterial and venous concentrations of unesterified fatty acids. J Lab Clin Med 1959;53:833–841.
15 Wagenmakers AJM, Coakley JH, Edwards RHT: Metabolism of branched-chain amino acids and ammonia during exercise: clues from McArdle's disease. Int J Sports Med 1990;11:S101–S113.
16 Wagenmakers AJM: Role of amino acids and ammonia in mechanisms of fatigue; in Marconnet P, Komi P, Saltin B (eds): Proc IVth Nice Symp on Exercise and Sport Biology. Med Sport Sci. Basel, Karger, to be published.
17 Siliprandi N: Transport and function of carnitine: Relevance to carnitine-deficient diseases. Ann NY Acad Sci 1986;488:118–125.
18 Siliprandi N: Carnitine in physical exercise; in Benzi G, Packer L, Siliprandi N (eds): Biochemical Aspects of Physical Exercise. Amsterdam, Elsevier, 1986, pp 197–206.
19 Cerretelli P, Marconi C: L-Carnitine supplementation in humans. The effects on physical performance. Int J Sports Med 1990;11:1–14.
20 Hülsmann WC, Siliprandi D, Ciman M, Siliprandi N: Effects of carnitine on the

oxidation of 2-oxoglutarate to succinate in the presence of acetoacetate or pyruvate. Biochim Biophys Acta 1964;93:166–168.

21 Van Hinsbergh VW, Veerkamp JW, Engelen PJM, Ghijsen WJ: Effect of L-carnitine on the oxidation of leucine and valine by rat skeletal muscle. Biochem Med 1978;20: 115–124.
22 Paul HS, Adibi SA: Effect of carnitine on branched chain amino acid oxidation by liver and skeletal muscle. Am J Physiol 1978;234:E494–E499.
23 Lennon DFL, Stratman FW, Shrago E, Nagle FJ, Madden M, Hanson P, Carter AL: Effects of acute moderate-intensity exercise on carnitine metabolism in men and women. J Appl Physiol 1983;55:489–495.
24 Carlin JI, Reddan WG, Sanjak M, Hodach R: Carnitine metabolism during prolonged exercise and recovery in humans. J Appl Physiol 1986;61:1275–1278.
25 Hiatt WR, Regensteiner JG, Wolfel EE, Ruff L, Brass EP: Carnitine and acylcarnitine metabolism during exercise in humans. J Clin Invest 1989;84:1167–1173.
26 Janssen GME, Scholte HR, Vaandrager-Verduin MHM, Ross JD: Muscle carnitine level in endurance training and running a marathon. Int J Sports Med 1989;10(suppl 3):S153–S155.
27 Sjøgaard G, Saltin B: Extra- and intracellular water spaces in muscles of man at rest and with dynamic exercise. Am J Physiol 1982;243:R271–R280.
28 Sejersted OM, Vøllestad NK, Medbø JI: Muscle fluid and electrolyte balance during and following exercise. Acta Physiol Scand 1986;128(suppl 556):119–127.
29 Harris RC, Foster CV, Hültman E: Acetylcarnitine formation during intense muscular contraction in humans. J Appl Physiol 1987;63:440–442.
30 Soop M, Björkman O, Cederblad G, Hagenfeldt L, Wahren J: Influence of carnitine supplementation on muscle substrate and carnitine metabolism during exercise. J Appl Physiol 1988;64:2394–2399.
31 Marconi C, Sassi G, Carpinelli A, Cerretelli P: Effects of L-carnitine loading on the aerobic and anaerobic performance of endurance athletes. Eur J Appl Physiol 1985; 54:131–135.
32 Borum PR: Plasma carnitine compartment and red blood cell carnitine compartment of healthy adults. Am J Clin Nutr 1987;46:437–441.
33 Cooper MB, Jones DA, Edwards RHT, Corbucci GC, Montanari G, Trevisani C: The effect of marathon running on carnitine metabolism and on some aspects of muscle mitochondrial activities and antioxidant mechanisms. J Sports Sci 1986;4: 79–87.
34 Suzuki M, Kanaya M, Muramatsu S, Takahashi T: Effects of carnitine administration, fasting, and exercise on urinary carnitine excretion in man. J Nutr Sci Vitaminol 1976;22:169–174.
35 Greig C, Finch KM, Jones DA, Cooper M, Sargeant AJ, Forte CA: The effect of oral supplementation with L-carnitine on maximum and submaximum exercise capacity. Eur J Appl Physiol 1987;5:457–460.
36 Siliprandi N, Di Lisa F, Pieralisi G, Ripari P, Maccari F, Menabo R, Giamberardino MA, Vecchiet L: Metabolic changes induced by maximal exercise in human subjects following L-carnitine administration. Biochim Biophys Acta 1990;1034:17–21.
37 Lennon DLF, Shrago ER, Madden M, Nagle FJ, Hanson P: Dietary carnitine intake related to skeletal muscle and plasma carnitine concentrations in adult men and women. Am J Clin Nutr 1986;43:234–238.

38 Dubelaar ML, Lucas CMHB, van der Veen FH, Wellens HJJ, Hülsmann WC: The acute effect of L-carnitine on skeletal muscle force tests in the dog. Pflügers Arch 1990;416:S2.
39 Paulson DJ, Shug AL: Tissue specific depletion of L-carnitine in rat heart and skeletal muscle by D-carnitine. Life Sci 1981;28:2931–2938.
40 Bazzato G, Coli U, Landini S, Mezzina C, Ciman M: Myasthenia-like syndrome after D,L- but not L-carnitine. Lancet 1981;i:1209.
41 Keith RE: Symptoms of carnitine-like deficiency in a trained runner taking D,L-carnitine supplements. JAMA 1986;255:1137.
42 Voeffray R, Perlberger JC, Tenud L, Gosteli J: L-Carnitine: Novel synthesis and determination of the optical purity. Helv Chim Acta 1987;70:2058–2064.
43 Van Hinsbergh VWM: Fatty acid and leucine oxidation in human and rat muscle; thesis University of Nijmegen 1979.

Dr. A.J.M. Wagenmakers, Department of Human Biology,
Nutrition Research Centre, University of Limburg, P.O. Box 616,
NL-6200 MD Maastricht (The Netherlands)

Aspects of Dehydration and Rehydration during Exercise

N.J. Rehrer

Department of Sports Medicine, Academic Hospital, Free University of Brussels, Belgium

Rationale for Fluid Ingestion

The primary goals of supplementation during endurance exercise are to maintain fluid balance and to enhance carbohydrate (CHO) availability and utilization. There is some debate as to which of these goals takes priority [3, 50]. However, because a negative fluid balance is associated with health risks [68] in addition to performance decreases [58], it is postulated that hydration, in most cases of endurance exercise, receives priority. There is no immediate health risk associated with limited CHO availability, although reduced CHO availability can limit performance. The advantages of supplementing with CHO immediately before and during exercise have been discussed in Chapter 1 of this volume. The present chapter will focus primarily on fluid provision and aspects of gastrointestinal function which influence this process.

Core body temperature rises during exercise as a result of the increased heat which is liberated when chemical energy is transformed into mechanical energy, as muscles contract. The higher the intensity of exercise is, the greater the increase in metabolic rate and heat production will be [45]. To compensate, circulation (to the periphery) is increased [30] and sweat production (cooling by evaporation) ensues. As the rate of energy conversion increases, i.e. exercise intensity increases, the sweat rate increases [24, 36, 45]. It is the water loss, primarily from sweating during intense endurance exercise, which can make the body's own stores of water limiting and can result in dehydration. A state of dehydration may be reached in which performance may be reduced due to a decrease in blood volume, stroke volume, circulatory capacity, and a decreased capacity to cool. Eventually, sweat production itself may become reduced, thereby further limiting cooling capacity of the body.

In terms of performance, it is not always the case that provision of water takes priority over provision of CHO. Different conditions, the climate, exercise intensity and duration, and one's body composition, size and unique physiology, will determine the relative rates of fluid loss and endogenous CHO oxidation. These losses, in turn, will determine the relative needs of CHO and water.

Limiting Factors

Limitations placed by ingestion and gastrointestinal function determine how the aims of supplementation can best be achieved.

Ingestion
The first step to successful supplementation is assuring that a supplement is ingested. A supplement must be available and palatable for it to be ingested in quantities sufficient to fulfill the needs. Ingestion itself may also pose problems, depending on the type of exercise. It may be difficult to swallow the supplement, and to swallow the supplement without simultaneously swallowing air. Swallowing large quantities of air commonly results in eructation and may cause abdominal discomfort.

Gastric Emptying
The second step in the provision of a supplement is gastric emptying. Most absorption of CHO and water occurs in the intestines, although there is some flux of water across the stomach wall [59]. Ingestate must, thus, first be emptied from the stomach before quantitative absorption can occur.

Numerous factors influence the gastric emptying rate [for review, see 8, 43]. One of these is the particle size of the ingestate. The larger the particle size, the slower the gastric emptying rate is. For this reason, liquids are recommended for nutritional supplementation during exercise, even when carbohydrate supplementation is the prime objective. Also, fluids are normally used since supplementation is, in most situations, undertaken to provide both water and CHO. Since fluids are provided during exercise to replace losses incurred during the exercise session, the quicker the supplementation is absorbed the better. It is acknowledged that by increasing the gastric emptying rate, the rate at which fluids and solutes are *available* for absorption will be increased and it is assumed that the rate of absorption and utilization will be increased. It has also been suggested that an accelerated gastric emptying will give less gastric discomfort which may occur due to increased pressure of a full stomach combined with the motion of certain types of exercise.

Effects of exercise on gastric emptying have been studied for nearly two centuries. In most cases there has not been a systematic approach to the experimentation with respect to type and intensity of exercise, type of meal and method used to monitor gastric emptying. This may explain the contradictory results which have been obtained. Early observations [4, 10, 25] led to the conclusion that 'gentle' (i.e. low intensity) exercise increases the gastric emptying rate and that 'strenuous' (i.e. high intensity) exercise reduces the gastric emptying rate. The cut-off points between low intensity (stimulatory effect) and high intensity (inhibitory effect) and a possible region of no effect were not clearly defined.

Interpretation of Gastric Emptying Data

The methods used to assess the exercise intensity often are of a rather uncontrolled nature, or vary greatly, making comparisons between different experiments with 'similar' intensities difficult. Different methods to measure the gastric emptying rate may also give different results. When gastric residue volumes are compared or if only drink volumes are compared (not including gastric secretion) different conclusions may be drawn. Additionally, when gastric (or beverage) residue is measured at only one time, the varying length of time between ingestion and measurement of the residue in different studies may give different results. Furthermore, emptying rate until one specific time may not be representative of the rate averaged over the total time taken to empty completely. Some researchers approach this problem by doing several tests for each treatment on each individual; stopping after an increasing length of time in each successive trial. Here the inter-trial variability may disguise treatment effects.

A series of controlled studies was recently conducted in which the double sampling technique of George [22] adapted by Beckers et al. [5] was used. This technique has the advantage of allowing for the measurement of the volume of beverage residue (and gastric secretion) serially, at several time points after ingestion of a beverage. Using this technique, which gives emptying curves, as are displayed in figure 1, experiments have been car-

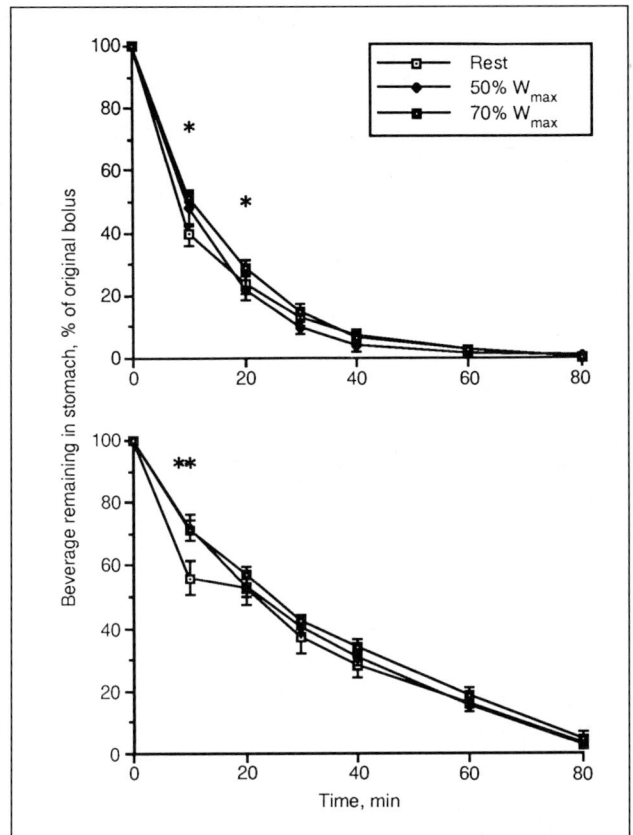

Fig. 1. Effects of exercise on gastric emptying of a 8 ml·kg body weight^{-1} bolus (~ 600 ml) of (*a*) 7%, primarily sucrose, CHO- and electrolyte-containing beverage (296 mosm·l^{-1}) and (*b*) an 18%, primarily maltodextrin, CHO- and electrolyte-containing beverage (444 mosm·l^{-1}). Significance of differences of remaining beverage volumes at specific times is given by * ($p < 0.05$) and ** ($p < 0.01$) [Data from 54].

ried out to study the effects of type and intensity of exercise on gastric emptying. Figure 1 shows the results of a bicycling study in which gastric emptying of two CHO- (7 and 18%) and electrolyte-containing beverages was monitored. One large bolus (8 ml·kg body weight^{-1}) was ingested, and measurements were made at different relative intensities, in identical tests, in the same individuals [54]. At exercise intensities of 50 and 70% of the maximal workload (W_{max}) no significant effect of exercise on gastric emptying of the CHO-containing beverages, relative to rest, was observed when the entire emptying curves were compared. There were, however, several time points at which the effects of exercise intensity were significant when a paired analysis was conducted. This example shows how the sampling time and statistical analysis can influence results. These findings are supported by results of Fordtran and Saltin [20], Costill and Saltin [14], and Feldman and Nixon [19] who also observed no significant effect at exercise intensities $\leq 70\%$ V_{O_2max} and a significant inhibitory effect above this level [14]. In contrast, Neufer et al. [46] found a stimulatory effect on gastric emptying with both 50 and 70% V_{O_2max} exercise. Neufer's singular results may be attributed to the fact that running was the mode of exercise used and in most of the other studies cycling was used. However, Fordtran and Saltin's [19] study was also performed with running exercise. Furthermore, both Rehrer et al. [55] and Houmard et al. [26] observed no differences in gastric emptying rates when the same individuals performed similar tests running and cycling.

Training Status

No difference was observed between trained bicyclists and untrained individuals in emptying 8 ml·kg body weight^{-1} (~ 600 ml) of different CHO-electrolyte-containing beverages or water at rest or during exercise [52]. These results are in opposition to those of Carrio et al. [11] who found that well-trained marathon runners had faster gastric emptying rates than untrained individuals at rest. One possible explanation for the contradictory results is that in the study of Carrio et al. [11], emptying of a solid meal was monitored, and in our study the emptying of fluids was measured. One can speculate that motility of the trained runners was greater in response to a solid meal. Alterations in motility may influence the emptying of fluids and solids differently. The method of measuring gastric emptying was also different in these two studies. Carrio et al. [11] used a meal labeled with technetium and a gamma camera to monitor the emptying of the meal, whereas the double sampling technique, referred to earlier in this chapter, was used in our work. However, in a recent study in which the emptying of fluids was measured simultaneously with the gamma camera and double sampling techniques, similar results were obtained [6].

Other factors which are known to influence the gastric emptying rate include: CHO composition, osmolality of the beverage ingested and the volume of beverage ingested.

CHO Type

In most early studies CHO concentration and osmolality were increased simultaneously. With the development of maltodextrin (glucose polymer) powders, which are readily dissolved in water, it has more recently become possible to develop solutions of similar CHO concentration but with different osmolalities. A high molecular weight glucose polymer solution of a given concentration has a lower osmolality than a mono- or disaccharide solution, or mixtures, of the same concentration. Again, contradictory results have been obtained when the gastric emptying rate of similarly concentrated glucose polymer solutions and glucose solutions are evaluated. Foster et al. [21] observed increased gastric emptying of a glucose polymer solution (75 mosm·l^{-1}), relative to a free glucose solution (266 mosm·l^{-1}), when residual gastric volumes were compared 30 min after ingestion. However, the difference in gastric emptying rate was calculated from *total* gastric residue volumes, including gastric secretions. When only beverage volumes were compared no significant difference was observed. Sole and Noakes [63] also compared gastric residues 15 min after ingestion of free glucose (739 mosm·l^{-1}) and glucose polymer (117 mosm·l^{-1}) solutions. They did observe a relatively faster gastric emptying of the polymer solution when only remaining beverage was compared, however not as great of a difference was observed as one would predict based upon the osmolality differences. When two solutions of similar glucose content but varying in osmolality (444 and 1,060 mosm·l^{-1}) were tested using the double sampling technique, only slight differences were observed [54]. As Sole and Noakes [63] noted, the magnitude of differences in gastric emptying rate are not in line with the magnitude of the differences in osmolality.

The effects of osmolality on gastrointestinal secretions, however, may be more important with respect to rehydration than the effects on gastric emptying. The importance of the observation made by Foster et al. [21], that a glucose solution elicited more gastric secretions than an isoenergetic glucose polymer solution has been neglected. A similar trend for increased gastric secretion with a glucose solution versus a polymer solution was also observed when the double sampling technique was applied [54]. Further supporting these findings is the observation made by Sole and Noakes [63] that there is a significant correlation between the volume of gastric secretion and the osmolality of the gastric residue.

The effects of increased osmolality of a solution to stimulate intestinal secretions are even more dramatic. Recent work utilizing intestinal perfu-

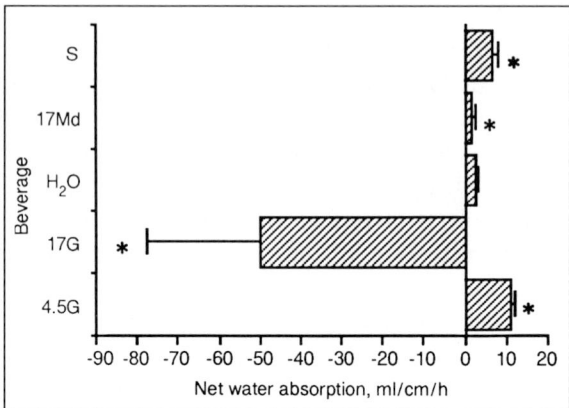

Fig. 2. Net water absorption or secretion (negative value) after intestinal perfusion with either S, a 6.1%, primarily sucrose, CHO- and electrolyte-containing beverage (296 mosm·l^{-1}) or 17Md, a 17% maltodextrin solution with 20 mEq·l^{-1} NaCl (301 mosm·l^{-1}), H$_2$O, water, 17G, a 17% glucose solution with 20 mEq·l^{-1} NaCl (1,223 mosm·l^{-1}) and 4.5G, a 4.5% glucose solution with 20 mEq·l^{-1} NaCl (313 mosm·l^{-1}). * p < 0.05, significantly different from water.

sion [53] showed that net secretion into the intestinal lumen occurred with a 17% glucose (sodium-containing) solution whereas net water absorption occurred with a 17% maltodextrin (sodium-containing) solution (fig. 2). These results support the earlier work of Miller [39] and the more recent studies of Leiper and Maughan [31] and Gisolfi et al. [23] in which an increase in intestinal secretion was observed with the presence of hypertonic fluid in the intestine. The greater osmolality of the glucose solution apparently induced increased intestinal secretion. These facts may explain the gastrointestinal complaints frequently observed after hypertonic beverage ingestion, particularly with intensive endurance exercise in which fluid balance may already be compromised [37, 57] (for review, see Chapter 10 of this volume).

Gastric Volume

The increased pressure due to an increased volume of gastric contents also stimulates gastric secretions [27]. However, with respect to fluid balance, this effect may be counterbalanced by the stimulatory effect that increased gastric volume has on gastric emptying. Volume is known to have a large stimulatory effect on gastric emptying [27, 35]. This results in exponential emptying after a large bolus (600 ml), as has been observed in the recent studies of Rehrer et al. [54, 55] with both CHO-containing as well as non-CHO beverages. This type of emptying pattern has been previously described by Hunt and Spurrell [29]. The practical importance of this observation is often overlooked when attempts are made to maximize fluid provision during athletic endeavor. The advice given to most athletes is to take in small quantities of beverage frequently during exercise; however, if larger quantities were ingested a faster gastric emptying rate would result [for review, see 51]. It has been speculated that there is a limit to the volume effect, with a maximum rate of emptying with a volume of approximately 600 ml [14]. However, Hunt and Spurrell [29] found increasing rates with volumes of up to 750 ml. And more recently, Mitchell and Voss [42] observed increasing emptying rates with volumes of up to 23 ml· $kg^{-1} \cdot h^{-1}$ which resulted in a gastric residue of over 1,000 ml. It may be that an adaptation or training effect on the emptying of large volumes occurs [C. Foster, pers. commun.]. This may explain the difference between the results of Costill and Saltin [14] and Mitchell and Voss [42].

CHO Concentration

The inhibitory effect of CHO concentration on gastric emptying is indisputable, although the effect is not so straight-forward as McHugh and Moran [34] and Brener et al. [7] have described. By using glucose solutions within a limited range of concentrations and volumes, it has been concluded that the gastric emptying rate is delayed so as to always provide a constant delivery of CHO to the intestines, in a 'steady-state condition'. With different forms of CHO, different volumes and concentrations, it has since been shown that a 'steady state' of CHO delivery is oversimplified. It is, rather, a continuum, in which the glucose concentration serves to delay emptying, modulated primarily by gastric volume. Fructose and glucose polymer solutions apparently do not elicit the same linear delay in gastric emptying with increasing concentration as glucose [63]. These results, as well as results of ours [54], have shown that with varying types of CHO solutions, the rate of energetic passage through the pylorus varies greatly.

Iso- or hypotonic solutions with a low CHO concentration ($\leq 7\%$), however, when given as one large bolus of ≥ 400 ml or as one large bolus followed by repeated drinking of boli of ~ 200 ml, empty at a similar rate as water [53, 54, 60]. Some researchers have observed differences with a 5% CHO solution versus water [46, 63]; however, these discrepancies may

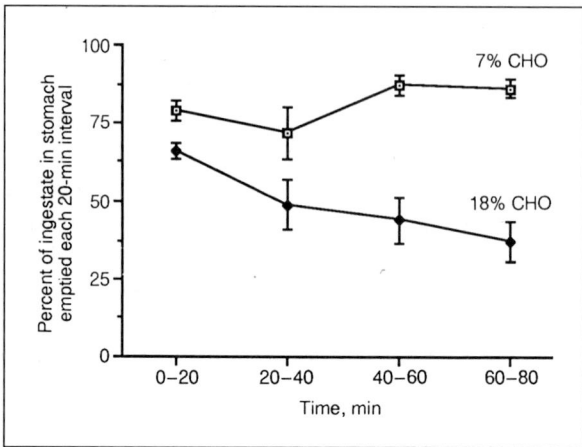

Fig. 3. Relative rates of gastric emptying with repeated drinking (8 ml·kg body weight^{-1} at 0 min and 2 ml·kg body weight^{-1} at 20, 40 and 60 min) of a 7% and a 18% CHO solution, taking into account the varying amount of volume remaining in the stomach. Amount emptied is expressed as a percentage of the volume of ingestate present in the stomach at the start of each 20-min interval. The percentage emptied was significantly decreased over time with the 18% solution but not with the 17% solution ($p < 0.05$) [data from 55].

be explained by varying time of sampling in the different studies. However, when a curve is based on multiple sampling, no significant differences are observed [54]. With smaller volumes (200 ml), when continuous sampling is conducted with gamma scintigraphy, water is observed to empty faster than a 5% glucose solution [38]. Apparently, the stimulatory effect of volume is reduced and the inhibitory effect of glucose content takes precedence.

A similar trade-off between the effects of volume and CHO content can be observed when gastric emptying rate is measured during repeated drinking (8 ml·kg body weight^{-1} followed every 20 min by 2 ml·kg^{-1}) of two different CHO-containing solutions (7 and 18%) [56]. With a 7% CHO solution, with repeated drinking, a high rate of gastric emptying was maintained, similar to that obtained during the first 20 min (the fast phase) after one large, single bolus. With an 18% solution, the emptying rate was reduced after an initial fast phase (fig. 3).

In general, as the CHO concentration of a solution increases, the gastric emptying rate decreases. However, with one bolus ingestion of 400–

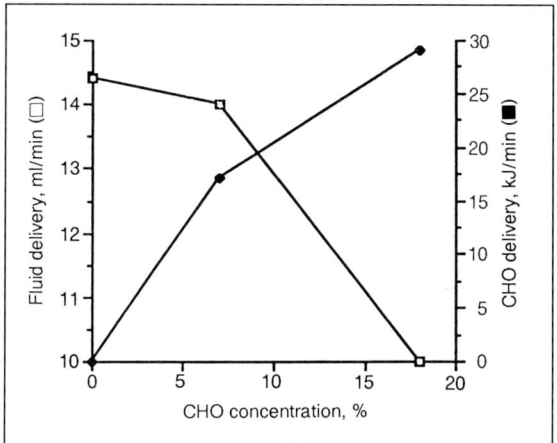

Fig. 4. Gastric emptying rates of CHO and fluid with solutions of different CHO concentration [data from 54, graphic design from 21].

600 ml, as the CHO concentration of the solution increases, the rate of fluid delivery decreases, but the rate of CHO delivery increases, at least up to concentrations of 18% [40, 41, 54]. If gastric emptying is used as a measure of 'availability', one obtains a continuum of fluid and CHO (energy) provision as shown in figure 4. Figure 4 is a plot of data from Rehrer et al. [54] in which beverage volumes were measured at 10-min intervals after one bolus of 8 ml·kg body weight^{-1} (~600 ml) was given and emptying rates were averaged over the first 40 min, since with some of the beverages the stomach was empty at this time. If the gastric emptying rate is limiting to the provision of fluid and CHO during exercise as Costill and Saltin [14] have suggested, one may draw the following conclusions from these data. If fluid losses are such that a state of dehydration would be reached before glycogen reserves become depleted, a beverage with a low CHO concentration should be ingested, since it will provide more fluid per unit time than a more concentrated solution. However, if CHO stores may become depleted before fluid losses become significant, a more concentrated beverage should be ingested since it will provide more CHO per unit time. The first situation may occur when exercise intensity is low and takes place in a hot (and humid) environment, or when an individual's sweat response is extreme. A possible scenario for the second situation is when exercise is in a cold environment and intensity is high and the duration is

relatively long, such as during cross-country skiing or winter game sports. Another point to consider when formulating a beverage to suit the situation is availability of the beverage. If an unlimited supply of beverage is available and easily ingested during exercise, such as is the case with professional bicyclists, ingestion of a concentrated solution may be advantageous [9]. In this case, although fluid losses are large, the continuous supply of fluid from beverages available from the onset of exercise, allows for sufficient rehydration, so that CHO supply may become the limiting factor. Although gastric emptying rate is decreased with a concentrated solution, the fact that supplementation can begin early and continue throughout exercise will compensate. However, with concentrated solutions, the osmolality is of critical importance, since as the osmolality increases so do the gastrointestinal secretions. If the volume of gastrointestinal secretions is large, shifts in fluid compartments, away from circulation and tissues and into the lumen, reduce the effectiveness of supplementation. For this reason, glucose polymers may be specifically advantageous in the formulation of concentrated solutions.

Electrolytes

Besides the knowledge that sodium stimulates intestinal absorption of glucose, and in combination with glucose stimulates intestinal absorption of water, it has been shown that a moderate amount of sodium increases gastric emptying of water [28, 62]. There is some indication that larger amounts of potassium may inhibit gastric emptying and may cause nausea [15]. However, in a controlled study, with 28 mEq·l^{-1} of potassium being added to a standard 15% maltodextrin solution, no variation in the gastric emptying rate was observed versus a control 15% maltodextrin solution or a 15% solution which had 28 mEq·l^{-1} sodium [53].

No enhancement of intestinal absorption is obtained by inclusion of potassium in a glucose solution. The reasons for inclusion in a rehydration beverage are to replace losses and possibly to stimulate intracellular hydration [44, 48]. However, the losses in sweat are not nearly so large [13] as those which are observed when the dehydration is a result of diarrhea [16], in which case inclusion of potassium may be warranted.

Beverage Temperature

Studies concerning the effects of beverage temperature on gastric emptying give equivocal results. MacArthur and Feldman [33] observed similar gastric emptying rates of hot (58 °C), warm (37 °C) or cold (4 °C) coffee at rest. Sun et al. [64] observed that hot (50 °C) and cold (4 °C) orange juice both emptied initially more slowly than warm (37 °C) orange juice, however half-emptying times were not significantly different. In the only study conducted during exercise, an increased rate of gastric emptying was observed when a CHO beverage was ingested at 5 °C, compared to ingestion at 15, 25 and 35 °C [14]. So far, the effect of temperature on gastric emptying remains enigmatic.

Dehydration

Dehydration also may negatively influence gastrointestinal function such that fluid uptake during exercise, after a significant amount of body water is lost (4–5% body weight), may be less efficient than in a state of fluid balance. Gastric emptying rate has been shown to be reduced during exercise when a beverage is given after dehydration which results in a reduction of 4–5% of the body weight [47, 56]. Furthermore, gastrointestinal complaints are common when dehydration and exercise are combined (see Chapter 10). Whether the reduction in gastric emptying is a cause of, or is a result of the symptoms, possibly relating to blood flow to the gastrointestinal tract, can only be speculated on at this time.

Significance of Gastric Emptying Rate

A question remains as to if the gastric emptying rates actually reflect 'availability' of water or CHO. When large differences in CHO delivery are observed as a result of varying CHO concentration, a much smaller difference in the amount of exogenous CHO oxidized is observed. An experiment was conducted in which various solutions containing ^{13}C-labeled CHOs were ingested throughout 80 min of exercise (70% V_{O_2max}). The oxidation of these substrates could be measured by measuring the ^{13}C/^{12}C ratio in the breath CO_2 and by taking breath measurements for total CHO utilization. This method is valid for calculating exogenous CHO oxidation when the subjects have a chronically low intake of ^{13}C sugars and identical tests are conducted without ^{13}C substrates, and background ^{13}C levels are taken into account when making calculations. Within the first 60 min a mean of 1,294 ml was ingested and at 80 min the remaining beverage volume in the stomach was measured. The mean amounts of CHO ingested, emptied and oxidized are displayed in figure 5.

The increase in exogenous CHO oxidation over time was, however, significantly greater with the 17% solutions than with the 4.5% solution, although no difference in oxidation was observed between the 17% glucose

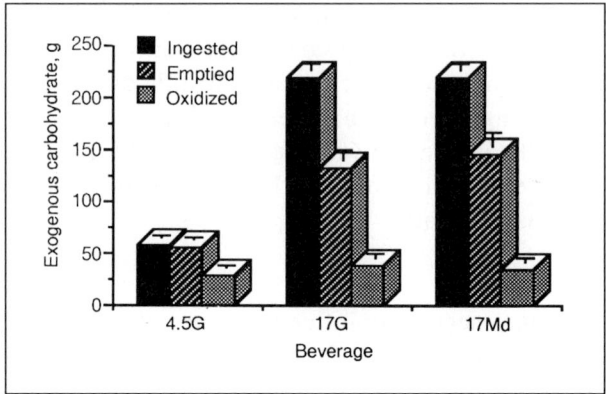

Fig. 5. Amounts of exogenous CHO ingested, emptied from the stomach and oxidized with ingestion of 1,294 ml of 4.5% (4.5G) and 17% glucose (17G) solutions and a 17% maltodextrin solution (17Md) during 80 min of 70% V_{O_2max} cycling exercise [Data from 53].

and 17% maltodextrin solutions (fig. 6). A similar, but somewhat greater difference was obtained by Wagenmakers et al. [65] when comparing oxidation of a 4% and a 16% solution over 2 h of 65% W_{max} (maximal working capacity in watts) exercise. Although the rates of utilization of CHO from a more concentrated solution are not as large one might expect based upon the rate of CHO delivery, the advantages of the increased rate of CHO delivery from a concentrated solution may only become apparent when endogenous reserves become limiting.

Another question arises concerning whether gastric emptying rates reflect water 'availability'. Various techniques have been used to measure intestinal absorption [for review, see 23, 32]. With the triple lumen technique an increase in net absorption of water in the intestine is observed with moderately concentrated carbohydrate-electrolyte solutions, which were found to empty from the stomach at the same rate as water [53, 54, Rehrer, unpubl. data (fig. 2)]. The results regarding intestinal absorption are in line with those of others [23, 31] and forms the basis for oral rehydration therapy in the clinical setting [18]. The enhanced intestinal absorption of water from a dilute CHO-sodium-containing solution can be explained by the coupled active transport of glucose and sodium in the intestine which simultaneously stimulates water absorption. This knowledge, regarding intestinal absorption, demon-

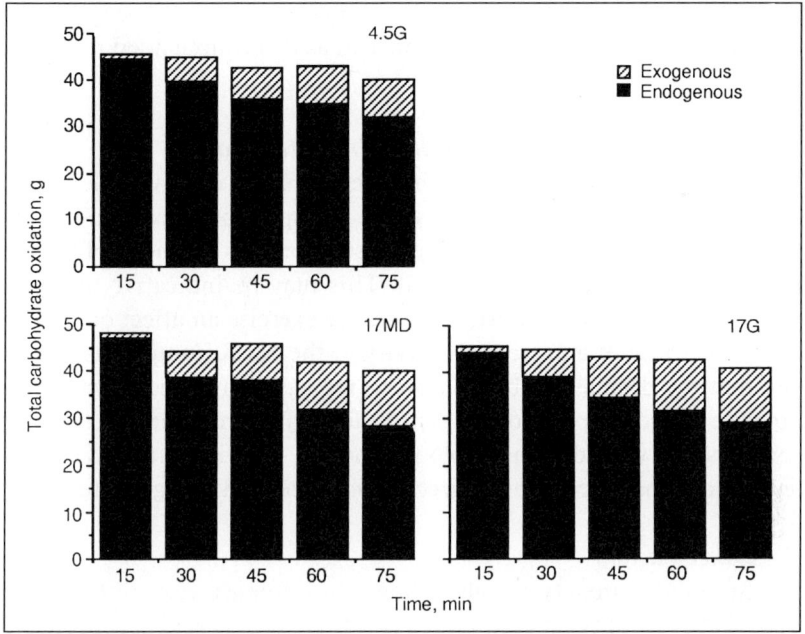

Fig. 6. Endogenous and exogenous CHO oxidation with ingestion of 1,294 ml of 4.5% (4.5G) and 17% glucose (17G) solutions and a 17% maltodextrin solution (17Md) during 80 min of 70% V_{O_2max} cycling exercise [Data from 53].

strates that not only gastric emptying significantly influences net fluid 'availability'.

Another point to consider with respect to (re)hydration is the retention of fluid and restoration of fluid volume. Nose et al. [52] have shown that restoration of total body water and plasma volume was greater with sodium and water than with water alone during rehydration following exercise-heat dehydration. These data, suggesting that water alone does not provide the most efficient restoration of body fluid pools, are supported by findings of Carter and Gisolfi [12] who found that during prolonged exercise (3 h) ad libitum consumption of a CHO-electrolyte beverage resulted in less of a plasma volume decrease than with water, although amounts ingested were not significantly different.

Noakes et al. [49] have reported numerous cases of hyponatremia when hypotonic fluids with low or no sodium are ingested in copious amounts during low intensity, ultra-endurance competition. This possible complication may well be avoided with ingestion of isotonic CHO-electrolyte fluids with a greater sodium content, as is recommended for oral rehydration solutions (35–60 mmol·l^{-1}) [17].

Energy Supply of the Gastrointestinal Membrane

Another unknown factor is the effect of CHO energy depletion on the gastrointestinal membrane. The active transport of glucose/sodium is dependent upon Na/K-ATPase. A decrease in protein synthesis of the gut has been observed during exercise [66]. This may be indicative of an energy deficit. It is unknown if during endurance exercise an affect on absorption occurs, because of a decreased activity in the Na/K pump. One can speculate that the decrease in blood flow to the gastrointestinal tract may compromise oxygen supply to the gastrointestinal membrane (and possibly in combination with decreased blood glucose) may inhibit active transport and absorption directly or indirectly via a reduced energy state.

Other Factors

Stress and anxiety are also known to influence gastrointestinal function. Gastric emptying is frequently slowed, gastric acid secretion is often observed to be increased and the motility of the gastrointestinal tract may be altered [for review, see 67]. A decrease in jejunal absorption has also been observed in relation to stress [1, 2].

The effects of stress may influence functioning of the gastrointestinal tract of competitive sportsmen and women more frequently than is generally acknowledged (see Chapter 10).

Conclusions

Obviously, from the above discussion, one cannot judge the efficacy of a (re)hydration regimen simply by measuring the gastric emptying rate. Exemplary of this is the fact that net intestinal absorption of water from two beverages may vary significantly and yet gastric emptying may proceed at similar rates. Additionally, it has not been demonstrated that by increasing the rate of gastric emptying, an increase in net intestinal absorption of solutes and water necessarily occurs. The absorptive capacity of the intestine will form an upper limit, above which an increase in gastric emptying rate will not influence absorption. In certain situations, an increase in the emptying rate above this upper limit may actually be counterproductive. Such is the case in 'dumping syndrome'. In this situation, CHO is supplied to the intestines faster than can be absorbed and osmotic diarrhea and/or fermentation in the colon results [61].

Selecting the composition and the manner of intake of a supplement during exercise should be based upon anticipated needs and knowledge of digestive function. This includes considering, in particular, CHO concentration, osmolality, electrolyte composition, volume, drinking pattern and palatability. Temperature of a beverage may be more important for the last of these reasons than for any special effect of the temperature on gastrointestinal function.

Timing of supplementation should be in anticipation of fluid and CHO needs for maximal benefit and a reduction of the risk of gastrointestinal dysfunction as a result of dehydration. Since individual variations in gastrointestinal function exist, any new supplementation regimen should be tested in a training session, similar to that experienced during competition or an important endurance event.

References

1 Barclay GR, Turnberg LA: Effect of psychological stress on salt and water transport. Gastroenterology 1987;93:91–97.
2 Barclay GR, Turnberg LA: Effect of cold-induced pain on salt and water transport in the human jejunum. Gastroenterology 1988;94:994–998.
3 Barr SI, Costill DL: Water: Can the endurance athlete get too much of a good thing? J Am Diet Assoc 1989;89:1629–1632, 1635.
4 Beaumont W: Experiments and Observations on the Gastric Juice and the Physiology of Digestion. Reprinted by Andrew Come, Edinburgh 1838.
5 Beckers EJ, Rehrer NJ, Brouns F, ten Hoor F, Saris WHM: Determination of total gastric volume, secretion, and residual meal using the double sampling technique of George. Gut 1988;29:1735–1729.
6 Beckers EJ, Leiper JB, Davidson J, Hemmel HG, Smith FW, Rehrer NJ, Brouns F, Saris WHM, ten Hoor F, Maughan RJ: A comparison of aspiration and scintigraphic techniques for measurement of gastric emptying rates of liquids in man. J Physiol 1990;429:107P.
7 Brener W, Hendrix TR, McHugh PR: Regulation of the gastric emptying of glucose. Gastroenterology 1983;85:76–82.
8 Brouns F, Saris WHM, Rehrer NJ: Abdominal complaints and gastrointestinal function during long-lasting exercise. Int J Sports Med 1987;8:175–189.
9 Brouns F, Saris WHM, Stroecken J, Beckers E, Thijssen R, Rehrer NJ, ten Hoor F: Eating, drinking and cycling, a controlled Tour de France simulation study. II. Effect of diet manipulation. Int J Sports Med. 1989;10(suppl 1):S41–S48.

10   Campbell JMH, Mitchell GO, Powell ATW: The influence of exercise on digestion. Guy's Hosp Rep 1928;78:279–293.
11   Carrio I, Estorch M, Serra-Grima R, Ginjaume M, Novitol R, Calabuig R, Vilardell F: Gastric emptying in marathon runners. Gut 1989;30:152–155.
12   Carter JE, Gisolfi CV: Fluid replacement during and after exercise in the heat. Med Sci Sports Exerc 1989;21:532–539.
13   Costill DL: Sweating: its composition and effects on body fluids; in Milvy P (ed): The Marathon: Physiological, Medical, Epidemiological and Psychological Studies. New York, New York Academy of Sciences, 1977, pp 161–174.
14   Costill DL, Saltin B: Factors limiting gastric emptying. J Appl Physiol 1974;37: 679–683.
15   Coyle EF, Costill DL, Fink WJ, Hoopes DG: Gastric emptying rates for selected athletic drinks. Res Q 1978;49:119–124.
16   Darrow DC, Pratt EL, Flett J, Gamble AH, Wiese HF: Distribution of water and electrolytes in infantile diarrhea. Pediatrics 1949;3:129–156.
17   Elliott EJ, Cunha-Ferreira R, Walker-Smith JA, Farthing MJG: Sodium content of oral rehydration solutions: at reappraisal. Gut 1989;30:1610–1621.
18   Farthing MJG: History and rationale for oral rehydration and recent developments in formulating an optimal solution. Drugs 1988;36(suppl 4):80–90.
19   Feldman M, Nixon JV: Effect of exercise on postprandial gastric secretion and emptying in humans. J Appl Physiol 1982;53:851–854.
20   Fordtran JS, Saltin B: Gastric emptying and intestinal absorption during prolonged severe exercise. J Appl Physiol 1967;23:331–335.
21   Foster C, Costill DL, Fink WJ: Gastric emptying characteristics of glucose and glucose polymer solutions. Res Q 1980;51:299–305.
22   George JD: New clinical method for measuring the rate of gastric emptying: the double sampling test meal. Gut 1968;9:237–242.
23   Gisolfi CV, Summers RV, Schedl HP, Bleiler TL, Oppliger RA: Human intestinal absorption: direct vs. indirect measurements. Am J Physiol 1990;258:G216–G222.
24   Greenhaff PL, Clough PJ: Predictors of sweat loss in man during prolonged exercise. Eur J Appl Physiol 1989;58:348–352.
25   Hellebrandt FA, Tepper RH: Studies on the influence of exercise on the digestive work of the stomach. Am J Physiol 1934;107:355–363.
26   Houmard JA, Egan PC, Johns RA, Neufer PD, Chenier TC, Israel RG: Gastric emptying during running and cycling (abstract). Med Sci Sports Exerc 1990; 22(suppl):S90.
27   Hunt JN, MacDonald I: The influence of volume on gastric emptying. J Physiol 1954;126:459–474.
28   Hunt JN, Pathak JD: The osmotic effects of some simple molecules and ions on gastric emptying. J Physiol 1960;154:254–269.
29   Hunt JN, Spurrell WR: The pattern of emptying of the human stomach. J Physiol 1951;113:157–168.
30   Johnson JM: Regulation of skin circulation during prolonged exercise; in Milvy P (ed): The Marathon: Physiological, Medical, Epidemiological and Psychological Studies. New York, New York Academy of Science, 1977, pp 195–212.
31   Leiper JB, Maughan RJ: Absorption of water and electrolytes from hypotonic, isotonic and hypertonic solutions. J Physiol 1986;373:90P.

32 Leiper JB, Maughan RJ: Experimental models for the investigation of water and solute transport in man. Drugs 1988;36(suppl 4): 65–79.
33 McArthur KE, Feldman M: Gastric acid secretion gastrin release and gastric emptying in humans as affected by liquid meal temperature. Am J Clin Nutr 1989;49: 51–54.
34 McHugh PR, Moran TH: Calories and gastric emptying: a regulatory capacity with implications for feeding. Am J Physiol 1979;236:R254–R260.
35 Marbaix O: Le passage pylorique. Cellule 1898;14:249–330.
36 Maughan RJ: Thermoregulation and fluid balance in marathon competition at low ambient temperatures. Int J Sports Med 1985;6:15–19.
37 Maughan RJ, Fenn CE, Leiper JB: Effects of fluid electrolyte and substrate ingestion on endurance capacity. Eur J Appl Physiol 1989;58:481–486.
38 Maughan RJ, Leiper JB: Studies in human models. Clin Ther 1990;12(suppl A): 63–72.
39 Miller TG: Intestinal intubation. Cleve Clin Q 1949;16:68.
40 Mitchell JB, Costill DL, Houmard JA, Flynn MG, Fink WJ, Beltz JD: Effects of carbohydrate ingestion on gastric emptying and exercise performance. Med Sci Sports Exerc 1988;20:110–115.
41 Mitchell JB, Costill DL, Houmard JA, Fink WJ, Rober025 RA, Davis JA: Gastric emptying: influence of prolonged exercise and carbohydrate concentration. Med Sci Sports Exerc 1989;21:269–274.
42 Mitchell JB, Voss KW: The influence of volume of fluid ingested on gastric emptying and body fluid balance (abstract). Med Sci Sports Exerc 1990;22(suppl):S90.
43 Murray R: The effects of consuming carbohydrate-electrolyte beverages on gastric emptying and fluid absorption during and following exercise. Sports Med 1987;4: 322–351.
44 Nadel ER, Mack GW, Nose H: Influence of fluid replacement beverages on body fluid homeostasis during exercise and recovery; in Gisolfi CV, Lamb DR (eds): Perspectives in Exercise, vol. 3: Fluid Homeostasis. Carmel, Ind, Benchmarck Press, 1990, pp 181–205.
45 Nadel ER, Wenger CB, Roberts MF, Stolwijk JAJ, Cafarelli E: Physiological defences against hyperthermia of exercise; in Milvy P (ed): The Marathon: Physiological, Medical, Epidemiological and Psychological Studies. New York, New York Academy of Science, 1977, pp 98–109.
46 Neufer PD, Costill DL, Fink WJ, Kirwan JP, Fielding RA, Flynn MG: Effects of exercise and carbohydrate composition on gastric emptying. Med Sci Sports Exerc 1986;18:658–662.
47 Neufer PD, Young AJ, Sawka MN: Gastric emptying during exercise: effects of heat stress and hypohydration. Eur J Appl Physiol 1989;58:433–439.
48 Nielsen B, Sjogaard G, Ugelvig J, Knudsen B, Dohlman B: Fluid balance in exercise dehydration and rehydration with different glucose-electrolyte drinks. Eur J Appl Physiol 1986;55:318–325.
49 Noakes TD, Goodwin N, Rayner BL, Branken T, Taylor RKN: Water intoxication: a possible complication during endurance exercise. Med Sci Sports Exerc 1985;17: 370–375.
50 Noakes TD: The dehydration myth and carbohydrate replacement during prolonged exercise. Cycling Sci 1990 (June):23–29.

51 Noakes TD, Rehrer NJ, Maughan RJ: The importance of volume in regulating gastric emptying. Med Sci Sports Exerc, in press.
52 Nose H, Mack GW, Xiangrong S, Nadel ER: Role of osmolality and plasma volume during rehydration in humans. J Appl Physiol 1988;65:325–331.
53 Rehrer NJ: Limits to fluid availability; PhD thesis, University of Limburg, Maastricht, The Netherlands, 1990.
54 Rehrer NJ, Beckers EJ, Brouns F, ten Hoor F, Saris WHM: Exercise and training effects on gastric emptying of carbohydrate beverages. Med Sci Sports Exerc 1989; 21:540–549.
55 Rehrer NJ, Brouns F, Beckers EJ, ten Hoor F, Saris WHM: Gastric emptying with repeated drinking during running and bicycling. Int J Sports Med 1990;11:238–243.
56 Rehrer NJ, Beckers EJ, Brouns F, ten Hoor F, Saris WHM: Effects of dehydration on gastric emptying and gastrointestinal distress while running. Med Sci Sports Exerc 1990;22:790–795.
57 Rehrer NJ, van Kemenade MC, Meester TA, Brouns F, Saris WHM: Gastrointestinal complaints in relation to dietary intakes in triathletes. Int J Sports Nutr, in press.
58 Saltin B: Aerobic work capacity and circulation at exercise in man. Acta Physiol Scand 1964;62(suppl 230):1–52.
59 Scholer JF, Code CF: Rate of absorption of water from stomach and small bowel of human beings. Gastroenterology 1954;27:565–577.
60 Seiple RS, Vivian VM, Fox EL, Bartels RL: Gastric-emptying characteristics of two glucose polymer-electrolyte solutions. Med Sci Sports Exerc 1983;15:366–369.
61 Sessions RT, Reynolds VH, Ferguson JL, Scott HW: Correlation between introduodenal osmotic pressure changes and ^{51}Cr blood volumes during induced dumping in men with normal stomachs. Surgery 1962;52:266–278.
62 Shay H, Gershon-Cohen J: Experimental studies of gastric physiology in man. II. A study of pyloric control: the roles of acid and alkali. Surg Gynecol Obstet 1934;58: 935–955.
63 Sole CC, Noakes TD: Faster gastric emptying for glucose-polymer and fructose solutions than for glucose in humans. Eur J Appl Physiol 1989;58:605–612.
64 Sun WM, Houghton LA, Read NW, Grundy DG, Johnson AG: Effect of meal temperature on gastric emptying of liquids in man. Gut 1988;29:302–305.
65 Wagenmakers AJM, Brouns F, Saris WHM, Halliday D: Maximal oxidation of oral carbohydrates during exercise (abstract). Med Sci Sports Exerc 1990;22(suppl): S120.
66 Wasserman D, Geer R, Williams P, Becker T, Lacy D, Abumrad N: Interaction of gut and liver in nitrogen metabolism during exercise. Metabolism, in press.
67 Wolf S: The psyche and the stomach. Gastroenterology 1981;80:605–614.
68 Wyndham CH: Heatstroke and hyperthermia in marathon runners; in Milvy P (ed): The Marathon: Physiological, Medical, Epidemiological and Psychological Studies. New York, New York Academy of Science, 1977, pp 129–138.

N.J. Rehrer, PhD, Department of Sports Medicine, Academic Hospital,
Free University of Brussels, Laarbeeklaan 101, B–1090 Brussels (Belgium)

High Intensity Exercise Performance and Acid-Base Balance: The Influence of Diet and Induced Metabolic Alkalosis

R.J.Maughan[a], P.L. Greenhaff[b]

[a] Department of Environmental and Occupational Medicine, University Medical School, Aberdeen, and
[b] School of Sport and Exercise Sciences, University of Birmingham, UK

Introduction

The ability to perform prolonged strenuous exercise is primarily determined by the capacity of the cardiovascular system to deliver oxygen to the working muscles and by the capacity of those muscles to produce energy by the oxidative metabolism of fat and carbohydrate (CHO). These factors, which determine an individual's maximum oxygen uptake (V_{O_2max}), set an upper limit to the rate at which energy can be made available by oxidative metabolism. For sedentary individuals, the V_{O_2max} is about 10 times the resting rate of oxygen consumption, whereas highly trained endurance athletes can achieve values about 20 times resting V_{O_2}. For short periods of time it is possible to exercise at workloads far higher than that at which V_{O_2max} is achieved, by relying on anaerobic metabolism to supply energy. In the transition from rest to maximum exercise, the rate of energy turnover in the exercising muscles can increase by as much as 1,000-fold. The duration of such activity is severely limited, however, as fatigue occurs rapidly.

The immediate source of energy for the contracting muscles is adenosine triphosphate (ATP), and studies on animal muscle in which the resynthesis of ATP is prevented have shown that the amount of work which can be performed is very small: only about three normal twitch contractions are possible [13]. However, the maximum activity of the enzyme creatine kinase, which transfers a phosphate group from creatine phosphate (CP) to adenosine diphosphate (ADP) resulting in the resynthesis of ATP, is higher than that of adenosine triphosphate (ATPase) which catalyses the breakdown of ATP to ADP [45]. This ensures that the ATP content of the muscle is kept high as long as there is adequate CP available: only when the CP content has fallen to less than half the resting value will a fall in the ATP level be apparent. Although a fall in ATP is observed only in intense exercise, the decline in muscle ATP concentration is, in itself, probably not the cause of failure of contraction [22, 86]. The fall in muscle ATP concentration reflects a decreasing rate of ADP rephosphorylation by CP and an increased rate of dephosphorylation of ADP to adenosine monophosphate (AMP).

When only a few muscle contractions are involved and the duration of the exercise is no more than 1–2 s, most (perhaps 80%) of the energy required will be provided from ATP and CP. In the post-exercise period, the muscle CP and ATP content will return to normal within a few minutes, with the energy for their resynthesis being derived from oxidative metabolism.

When the duration of the exercise is increased to a few (5–10) seconds, significant falls in the muscle ATP and CP content are observed. Although it was thought at one time that no lactate would be produced in exercise lasting less than 10 s, many studies have now shown that an increase in anaerobic glycosis occurs in muscle almost immediately at the onset of high intensity exercise. With electrical stimulation of the quadriceps, lactate accumulation occurs within 5 s of the initiation of muscle activation [44]. Sprinting over a distance of 40 m (in about 5 s) has also been shown to cause a large increase in the lactate content of the quadriceps muscles [39]. In a maximal cycling task of 30 s duration, Boobis [8] observed that CP breakdown and lactate formation by glycolysis contributed about equally to the total energy requirement during the first 6 s of exercise; he reported, however, that only about 35% of the muscle CP content was used in the first 6 s of exercise, and that a further substantial contribution was made over the next 24 s during which time the power output fell progressively. The fall in power output occurred in parallel with the fall in the contribution of CP to ATP resynthesis. The muscle ATP content, which was essentially unchanged after 6 s, had fallen by almost half at the end of the 30 s period. The muscle glycogen content had fallen by 16% of the resting value after 6 s and by 30% after 30 s. It must, of course, be remembered that these data refer to measurements made on whole muscle samples and may obscure changes occurring within individual muscle fibres.

It has long been known that the maximum speed which can be achieved by elite sprinters begins to decline after about 50 m, and in maximal cycle ergometer tests the peak power output is observed within 1–3 s of the onset of exercise. The reasons for this now become apparent. In the initial acceleration phase, most of the energy is derived from CP breakdown, and after only a few seconds the CP content of the muscle has fallen to the point where the rate of resynthesis of ATP by this mechanism cannot be sustained. Almost immediately during the exercise, the rate of anaerobic glycolysis is dramatically increased but the maximum rate of ATP resynthesis by glycolysis is still less than can be achieved by transfer of phosphate groups from CP so long as the muscle CP content remains high. As exercise continues, the rate of glycolysis also begins to fall resulting in a further decline in the rate of ATP production [85].

In most field games such as soccer, the pattern of activity consists of sprints lasting no more than a few seconds followed by recovery. During these high intensity bursts, most of the energy will be supplied by CP, with glycolosis contributing more as the distance increases. The resynthesis of CP after exercise has been reported to be complete within a few minutes after work [32, 41]. Where greater demands are placed on anaerobic metabolism, however, this process may be delayed, as it has been shown that the CP content of muscle did not return to the pre-exercise level even 60 min after high intensity exercise where exhaustion was reached after about 3 min [4]. If a second sprint is performed before CP levels have been restored, the speed and duration may be reduced. Restoration of CP levels during recovery will not be delayed if low intensity exercise is performed rather than complete rest.

As the exercise becomes more prolonged, the contribution of glycolysis to energy production is increased leading to the accumulation of lactate in the muscle and blood. In exercise lasting from about 10 s to about 2–3 min, anaerobic glycolysis will be the main source of energy. In longer events, oxidative metabolism gradually becomes more important: when the exercise duration exceeds about 3 min, aerobic processes supply more than half the total energy requirement [4]. In the shorter events, most of the fuel used will be in the form of muscle glycogen, but glucose derived from the blood becomes more important as the duration increases.

Muscle Buffering and Fatigue

At the pH which exists within muscle, the lactic acid which is formed by glycolysis is almost totally dissociated to form the negatively charged lactate anion and a proton (a positively charged hydrogen ion, H^+). Some of these protons are buffered within the muscle and some leave the muscle, but the high rates of anaerobic glycolysis which occur during intense exercise result in a marked acidosis within the active muscle cells. Muscle pH, however, does not normally decline below about 6.5 even during exhausting high intensity exercise [77]: this demonstrates the effectiveness of the available muscle buffering mechanisms, as the quantity of hydrogen ions produced during such exercise would, if not neutralized, cause the intramuscular pH to fall to about 2.

Buffering capacity was first defined by van Slyke in 1922 as the amount of free hydrogen or hydroxyl ions required to change a solution by 1 pH unit. The buffer mechanisms in the mammalian organism can be conveniently classified into intracellular and extracellular systems. Briefly, the intracellular buffering system comprises physicochemical (static) and metabolic buffer mechanisms, contributing about 60 and 40% respectively to the intracellular buffering capacity. Hultman and Sahlin [43] have published a detailed review of these mechanisms. Some of the hydrogen ions produced during exercise leave the muscle and enter the extracellular space, where they encounter the extracellular buffering systems, of which the most effective are bicarbonate and haemoglobin. (For a detailed review of these mechanisms, see Siggaard-Andersen [83]).

Whatever the buffering mechanism or the degree of buffering that it affords, hydrogen ions produced during intense exercise will eventually saturate the body's capacity to neutralize them, resulting in the development of an acute metabolic acidosis. Circumstantial evidence suggests that the accumulation of hydrogen ions within the muscle is closely related to the fatigue process, and may indeed be the cause [78]. A fall in the intracellular pH will affect many metabolic processes in the muscle. Among a number of inhibitory mechanisms, hydrogen ions are known to inhibit calcium release from the sarcoplasmic reticulum [64], to reduce the activity of myofibrillar ATPase [17], and to directly inhibit the interaction between actin and myosin [12]. The activity of a number of enzymes is pH sensitive; at low pH, both phosphorylase and phosphofructokinase will be inhibited, leading to a reduction in the rate of glycolysis and hence a reduction in the rate of ATP production [15, 65]. A low pH may also reduce the rate of ADP rephosphorylation by altering the creatine kinase equilibrium [33, 67] and will activate AMP deaminase [18].

Although the hydrogen ions produced by anaerobic glycolysis thus make a major contribution to the fatigue process, glycolysis permits a high rate of ATP formation to be sustained: a reduction in the rate of glycolytic flux would have the effect of reducing the rate of ATP resynthesis, and consequently the exercise intensity that could be sustained. Any mechanism that allowed an increased rate of glycolysis would allow a higher rate of work to be performed for a short period of time, although it would accelerate the fatigue process. Similarly, any mechanism capable of increasing muscle buffering capacity or resulting in an enhanced rate of hydrogen ion efflux from the muscle, has a potential for increasing exercise performance in situations where metabolic acidosis is the limiting factor. Many different approaches to the attainment of these aims have been attempted, both in the laboratory and by competitive sportsmen. Two of these, by nutritional intervention, will be discussed here: the manipulation of the pre-event diet and the alteration of acid-base status by the ingestion of alkalinizing agents.

Effects of Diet on Exercise Capacity

It has long been recognized that the capacity to perform prolonged exercise is closely related to the pre-exercise muscle glycogen content [6]. During prolonged exercise, a progressive utilization of the muscle glycogen stores takes place, and exhaustion coincides with the point at which the intramuscular glycogen content is depleted [37]. In high intensity exercise, however, the rate of energy provision by glycogenolysis is not normally limited by the availability of glycogen within the muscle. In exercise which results in exhaustion within a few minutes, large amounts of glycogen remain within the muscle at the point of fatigue, even though the rate of glycogenolysis is high during the exercise period [36, 82].

Perhaps for this reason, the possibility that dietary manipulations which alter the muscle glycogen store might affect the performance of brief, high intensity exercise have received relatively little attention. In 1975, however, Kelman et al. [50] reported that the blood lactate concentration was lower than normal after a low CHO diet and higher than normal after a high CHO diet during 5 min exercise bouts at workloads from 30 to 95% of V_{O_2max}. It was later shown that the endurance time in high intensity exercise (104% of V_{O_2max}) was also influenced by the preceding diet [62]. When subjects performed a high intensity exercise test after 3 days on a low CHO diet preceded by prolonged exhausting exercise (65 min at 75% V_{O_2max}) to deplete muscle glycogen, mean exercise time was reduced from 4.87 to 3.32 min (p < 0.005). This was followed by a high

CHO diet for 3 days and a further high intensity exercise test; exercise time after the high CHO diet was increased ($p < 0.05$) to 6.65 min. The diets used in this study were approximately isoenergetic, although there was a tendency for an increased energy intake on the low CHO diet and a decreased energy intake on the high CHO diet; these effects appeared to be related to the subjective sensations of satiety on the different diets. In terms of composition, the experimental diets were extreme: on the normal diet, CHO accounted for $43 \pm 9\%$ (means $\pm$ SD) of total energy intake; this was reduced to $3 \pm 2\%$ on the low CHO diet and increased to $84 \pm 6\%$ on the high CHO diet.

This study also confirmed the earlier observations regarding the effects of diet on the blood lactate concentration after exercise. The peak blood lactate concentration after exercise on the low CHO diet was lower than on the normal diet whereas blood lactate after the high CHO diet was higher than normal. Similar results were reported by Klausen and Sjogaard [53], who reported that muscle lactate concentration was lower during exercise at 106% of $V_{O_{2max}}$ when the pre-exercise muscle glycogen content was reduced by a low CHO diet compared with a high CHO diet condition. Richter and Galbo [72] later used an isolated rat muscle preparation to show that the rate of glycogenolysis during exercise was related to the muscle glycogen content. In a human model involving electrical stimulation of muscle, Hultman and Sjoholm [44] have demonstrated a decreased rate of glycogenolysis when the muscle glycogen content was low. Other results from electrical stimulation of human muscle, however, have indicated that the rate of glycogenolysis during 60 s of exercise was not different even when there were large differences in the initial glycogen content of the muscle [71].

The apparent dissociation between lactate accumulation and fatigue when exercise is performed after a low CHO diet is complicated by the differences in exercise duration on the different diets, but does indicate that the relationship between the accumulation of blood lactate and the subjective sensation of fatigue is strongly influenced by the preceding diet. Later studies, however, showed that the post-exercise blood lactate concentration was lower than normal after a low CHO diet and higher than normal after a high CHO diet even when the exercise intensity and duration were the same on all three dietary conditions [28].

This basic experimental design has since been repeated on a number of occasions [25–27]. Endurance time in cycle exercise at about 100% $V_{O_{2max}}$ is consistently and significantly reduced if the exercise test is preceded by a period of 3–4 days during which the CHO content of the diet is reduced to about 5–10% of total enery intake. This effect is observed even when the low CHO diet is not preceded by prolonged exhausting exercise intended to deplete the muscle glycogen stores. Feeding of a high CHO diet for 3–4 days has a less reproducible effect, although there is generally an increased endurance time. In these later studies, the energy intakes were more closely controlled, but the variations in composition were less extreme, with CHO accounting for about 65–75% of total energy intake on the high CHO phase of the diets. In the earlier study, where a significant improvement in endurance time was observed after the high CHO diet, the energy intake was less well controlled, but the composition of the diet was more extreme, with CHO accounting for 84% of the total energy intake [62].

The effect of fasting on the performance of high intensity exercise is similar to that of a low CHO diet. A 24-hour fast has been reported to reduce mean endurance time at a workload equivalent to 100% of $V_{O_{2max}}$ from 243 to 217 s [23].

In contrast to these effects on cycling exercise, there appears to be no effect of depletion of muscle glycogen by a combination of prolonged exercise and a low CHO diet on the performance of a muscle fatigue test consisting of 50 consecutive maximal unilateral isokinetic ($180° \cdot s^{-1}$) leg extensions [92]. Neither the peak torque nor the rate of decline of peak torque was different between the normal diet and low CHO diet conditions. The total duration of the exercise task in this study was not disclosed, but appears to be of the order of 40 s. Rather oddly, there was no decrease in the glycogen content of the exercising muscles during the exercise task, even though the blood lactate concentration rose to values in excess of 4 mmol/l. In contrast to this result, the performance of sustained isometric exercise by the quadriceps muscle at a force corresponding to 60% of maximum appears to be reduced if preceded by a prolonged exercise/low CHO diet regimen: the mean exercise duration in this study was 54 s on the normal diet condition and 46 s on the low CHO diet condition [61].

Diet, Muscle Glycogen and High Intensity Exercise

The normal glycogen content of resting human skeletal muscle is rather variable between individuals, although it remains rather constant in the absence of strenuous activity. Hultman [41] reported that the mean value for the glycogen content of the quadriceps muscle was about 86 mmol glucosyl units/kg, with a range of values from 63 to 133 mmol/kg. Bergstrom et al. [6] found a mean value of 108 mmol/kg. There is some debate as to whether there are differences in glycogen content between the different fibre types. A number of studies have shown differences between fibre type in glycogen storage in animal muscle, and there are some studies showing slightly higher values in type 2 fibres in human muscle [for review, see 1], but Essen and Henriksson [20], Blomstrand [7] and Fridén [21] among others have all reported that the glycogen content of type 1 and type 2 fibres is the same in human muscle. More recently, Roedde et al. [75] reported higher muscle glycogen values (115 mmol/kg) in trained subjects compared with untrained individuals (92 mmol/kg): this finding, however, is almost certainly the result of a reduction in the subjects' training load and a high CHO intake during the days preceding the sampling procedure. From the values reported in the literature, it seems reasonable to accept a figure of 80–100 mmol/kg as a value for the normal resting muscle glycogen content of human skeletal muscle.

Although there seems to be some variability between individuals in the muscle glycogen content, Hultman [41] reported that only hard exercise or extreme dietary alterations resulted in a significant change in the amount of glycogen stored. In the absence of strenuous exercise, short-term fasting has little effect on muscle glycogen content, in contrast to liver glycogen which falls rapidly. Hultman [41] reported that even after 5 days of fasting in individuals carrying out their normal daily activities but avoiding hard exercise, muscle glycogen concentration fell by less than 50%. The effects of a low CHO diet are similar to those of total fasting.

The rate of glycogenolysis in skeletal muscle increases in an approximately exponential manner as the exercise intensity is increased, and reflects the muscle fibre recruitment pattern. At low exercise intensities, only type 1 fibres, which have a high capacity for oxidative metabolism of both fat and CHO, are recruited [81]. As the load on the muscle is increased, type 2a fibres are recruited in addition to the type 1 fibres which remain active; these fibres also have a high oxidative capacity, but also have a high capacity for glycolysis, leading to an increased reliance on CHO as a fuel. At high workloads, type 2b fibres begin to be active. In these fibres, the glycolytic activity is high relative to the oxidative capacity, leading to a high rate of anaerobic glycogenolysis; the rate of pyruvate formation by glycolysis in these fibres exceeds the rate at which it can enter the Krebs cycle and conversion to lactate is inevitable.

Where high intensity exercise is avoided, therefore, the muscle glycogen stores are conserved; the muscular effort involved in the normal daily activities of most individuals can be accomplished by the oxidative fibres, and the glycolytic fibres are not active. When strenuous exercise is performed, however, there is necessarily a high rate of glycogenolysis. During exercise at 100% of V_{O_2max}, the rate of glycogen breakdown has been reported to be about 11 mmol glucosyl units/kg/min [90]. Given a normal muscle glycogen content of about 90–100 mmol/kg, and assuming a constant rate of glycogen utilization during exercise (an assumption which is almost certainly not valid, as the rate of glycogenolysis will fall as the exercise progresses), it would appear that the muscle glycogen store should be adequate to sustain exercise for much longer than the 3–6 min which is commonly observed to be the limit of endurance at this exercise intensity.

These measurements of muscle glycogen content and the rate of glycogen breakdown have been made on whole muscle, and many obscure changes taking place within specific fibre types. In very high intensity exercise the glycogen content of the type 2b fibres will fall rapidly, and in some fibres there may be almost no glycogen even when the exercise duration is short and the overall muscle glycogen content remains high. If this occurs, type 2 fibre fatigue may be the cause of a decrease in performance. After 3–4 days on a low CHO diet, there is likely to be some decrease in the muscle glycogen content, even if no strenuous exercise is performed, and it is possible that substrate availability may limit exercise capacity. This decrease is most likely to occur in the type 1 fibres in spite of their relatively low glycogenolytic capacity, as normal daily acitivities do not place much demand on the type 2 fibres [19]. If exercise is performed during this period of restricted CHO intake, the glycogen content of the muscle will certainly be reduced. In particular, if the exercise intensity is high, the type 2b muscle fibres, which have a high capacity for glycogenolysis, may lose most of their glycogen. If this happens, sprinting ability will be seriously impaired.

Diet, Exercise and Acid-Base Status

The effects on extracellular acid-base status of acute alterations in diet composition have been investigated on a number of occasions. Feeding a diet which is low ($<10\%$) in CHO results in a mild metabolic acidosis;

feeding a high CHO (> 65%) diet results in a tendency towards the development of a metabolic alkalosis, although this effect is less consistent [25–29]. In these experiments, where the diets are isoenergetic, the low CHO diet is high in fat and protein: the fat and protein intakes are reduced on the high CHO diet. In addition to alterations in macronutrient intake, dietary acid-base intake will be different on the different diets, and will largely reflect the protein intake [57].

Changes in blood-base excess are commonly accepted as the major index of acid-base status in conditions of metabolic acidosis or alkalosis, and relate to changes in the plasma concentration of strong anions or cations [83]. An alternative approach has been suggested by Stewart [87, 88], who focussed attention on the components of the buffer base. The term strong ion difference (SID), analogous to the traditional concept of the anion gap, relates to the difference between the sum of all strong cations minus the sum of all strong anions. These include the major inorganic ions, and also organic ions including lactate, keto acids and free fatty acids together with nonvolatile weak acids, primarily in the form of the plasma proteins. The plasma pH and bicarbonate concentration depend on pCO_2, the SID and on the plasma protein concentration, each of which can vary independently.

Acute changes in dietary macronutrient intake do not change the plasma concentration of sodium or chloride, the main strong ions, but a low CHO, high protein diet will result in increases in the concentration of free fatty acids, 3-hydroxybutyrate and plasma proteins. The metabolic acidosis which accompanies this diet therefore appears to be a result of an increase in the organic acid component of the SID together with an increase in the nonvolatile weak acids. It is well established that increasing or decreasing the plasma protein concentration will result in an acidosis or alkalosis respectively [56, 76]. When dietary protein intake is increased, an increased rate of protein oxidation will occur, accompanied by an increased rate of hydrogen ion formation [66].

In view of the recognized role of acidosis in limiting the performance of high intensity exercise, it is tempting to ascribe the reduction in exercise capacity which accompanies a low CHO diet to the effects of the diet on acid-base status rather than to any effect on the muscle glycogen content. A recent report, however, has suggested that administration of sodium bicarbonate will cause an acute reversal of the metabolic acidosis after a low CHO diet, but will not result in restoration of the capacity to perform high intensity exercise [3]. Although more evidence is needed, this suggests that it is indeed an effect on muscle glycogen content, perhaps specific to the type 2 fibres, which causes the diet-induced reduction in exercise capacity.

Influence of Induced Alkalosis: Animal Muscle Studies

At physiological pH, lactic acid is virtually completely dissociated and is the major source of hydrogen ion production during intense exercise. A part of the lactate produced within the muscle diffuses out into the extracellular space; the rates of lactate and hydrogen ion efflux are similar, but the relationship between the transport of these two species is not yet clear [5, 9, 54, 68, 79]. Whatever the mechanism of transport, results from animal preparation studies show that a change in the extracellular bicarbonate concentration can influence hydrogen ion flux across the muscle membrane. Unlike changes in CO_2 tension, which are known to effect the intramuscular pH [2], it is generally considered that alterations in the extracellular bicarbonate concentration have no such effect; as the membrane is impermeable to bicarbonate, increasing the extracellular bicarbonate concentration should have little influence on the intracellular bicarbonate concentration or pH [48, 74, 93]. It has, however, been demonstrated that a change in the bicarbonate concentration of the perfusate supplying isolated heart or diaphragm preparations, with constant tissue pCO_2, will result in significant changes of intracellular acid base status [2, 89]. Whether a similar response occurs in skeletal muscle in vivo is not known. Whatever the effect on intracellular pH, the efflux of lactate and hydrogen ions from muscle during concentration is accelerated by an increase in the muscle to blood pH or bicarbonate gradient [38, 60]. Coincident with the increased efflux rates is an increased work capacity during electrical stimulation and an enhanced post-exercise recovery of twitch tension.

Influence of Induced Alkalosis on Exercise Performance

As early as 1932, an increased exercise capacity and an increased blood lactate concentration were observed when bicarbonate was administered prior to exercise [16]. In view of the implications for athletic performance, it is hardly surprising that a large number of studies investigating the effects of metabolic alkalosis induced by ingestion of sodium bicarbonate or sodium citrate on the performance of high intensity exercise have been carried out since that initial report was published. The results, however, are by no means consistent or conclusive. Heigenhauser and Jones [35] have recently presented a comprehensive review of these studies.

Several investigators have reported a decrease in perceived exertion [73, 91] or an increase in performance [10, 14, 24, 63, 90] during high intensity exercise after bicarbonate administration. Others, however, have shown no benefit of an induced metabolic alkalosis on perceived exertion [70] or performance [11, 40, 51, 52, 55, 69]. In one study designed to simulate athletic competition, trained middle distance runners were used as subjects and the exercise consisted of a simulated 800 m race: in the alkalotic condition, subjects ran almost 3 s faster than in the placebo or control trials [94]. The reason for these conflicting results is not altogether clear, but is probably due in part to variations in the intensity and duration of the exercise tests used, in the dosage of sodium bicarbonate administered, an in the time delay between bicarbonate administration and the beginning of the exercise test (i.e. in the degree of metabolic alkalosis induced).

Performance has been monitored over exercise durations ranging from 30 s [46] to 20 min [49], and during continuous [24, 52] incremental [54, 90] and intermittent dynamic exercise as well as during isometric exercise [63]. In most studies, a dose rate of 0.3 g of sodium bicarbonate or citrate per kilogram body weight has been employed to induce alkalosis, and this has usually been administered orally in solution or in capsule form. Such a dose rate has usually resulted in an increase of 4–5 mmol/l in the plasma buffer base 2–3 h after administration [14, 54, 58, 63]. Horswill et al. [40] suggested that a dose rate of less than 0.3 g/kg body weight might be ineffective in improving exercise performance, but McKenzie et al. [58] reported that a dose rate of 0.3 g/kg was no more effective than was one half of this dose. From a recent study on racehorses, it appears that, because of the more marked pre-exercise acid-base changes, an increased dose (0.6 g/kg) might be more effective than the dose of 0.3 g/kg normally employed in human studies [30]; it also appeared from this study that bicarbonate administration might have a greater effect on performance if the exercise was not performed until at least 3 h after ingestion. This may partly explain the lack of an effect on performance noted by Kelso et al. [51] who used a higher than normal dose of 0.4 g/kg, but allowed an interval of only 1 h between treatment ingestion and exercise; plasma bicarbonate was elevated by only 3 mmol/l prior to exercise.

There are, of course, potential problems associated with the use of increased doses of bicarbonate. Vomiting and diarrhoea are not infrequently reported as a result of ingestion of even relatively small doses of bicarbonate, and this may limit any attempt to improve athletic performance by this method, certainly among those individuals susceptible to gastrointestinal problems. Sodium citrate administration, which also results in an alkaline shift in the extracellular fluid, has also been reported to improve peak power and total work output in a 60 s exercise test, but without any adverse gastrointestinal symptoms [59].

Where an increase in performance after bicarbonate ingestion has been observed, it has been ascribed to an increased rate of hydrogen ion efflux from the exercising muscles, reducing the rate of fall of intracellular pH, and relieving the pH-mediated inhibition of phosphofructokinase [90]. The higher blood lactate levels after exercise associated with metabolic alkalosis, even when the exercise duration is the same, may therefore be indicative not only of a higher rate of lactate efflux, but also of an increased contribution of anaerobic glycolysis to energy production.

Associated with the development of fatigue during high intensity exercise is a decline in the muscle adenine nucleotide content [34, 80]. The extent of the fall in muscle ATP concentration which occurs during maximal exercise in man has been shown to approach 40% of the pre-exercise level [8, 42]; even greater losses of ATP (60%) have been reported at exhaustion in the horse [84]. There is evidence to suggest that an increase in hydrogen ion efflux during near maximum intensity exercise after bicarbonate administration may decrease the extent of muscle adenine nucleotide loss during exercise [31]. Whether this is due to a pH-mediated decrease in the activation of AMP deaminase or an increased rate of ADP rephosphorylation via glycolysis is not clear. Whatever the mechanism, however, it seems reasonable to suggest that bicarbonate administration prior to high intensity exercise will only enhance performance when the intensity and duration of the exercise are sufficient to result in significant muscle acidosis and adenine nucleotide loss.

Conclusion

There is no doubt that, in certain well-defined situations, both of the procedures described above can enhance the performance of high intensity exercise, although the mechanism involved remain obscure. The effects of a high CHO diet appear to show considerable interindividual variability, but it is clear that a low CHO diet in the days before a high intensity exercise task will impair performance, either as a result of the metabolic acidosis which occurs or as a result of an inadequate muscle glycogen content. Many athletes do not allow sufficient rest between training and competition to allow glycogen resynthesis, nor do athletes engaged in short-duration events appear to perceive the need for an adequate (i.e. high) dietary CHO intake. Ingestion of large (0.3 g/kg body weight) amounts of sodium bicarbonate is contrary to the doping regulations governing most sports. Nonetheless, it does have potential for the improvement of short-term high intensity exercise, although the athlete is more likely to be deterred by the possibility of serious gastrointestinal problems than by ethical considerations.

References

1 Abernethy PJ, Thayer R, Taylor AW: Acute and chronic responses of skeletal muscle to endurance and sprint exercise; a review. Sports Med 1990;10:365–389.
2 Adler S, Arlene R, Relman R: Intracellular acid-base regulation. I. The response of muscle cells to changes in CO_2 tension and extracellular bicarbonate concentration. J Clin Invest 1965;44:8–20.
3 Ball D, Greenhaff PL, Maughan RJ: Dietary manipulation, sodium bicarbonate ingestion and the metabolic response to high-intensity exercise in man. J Physiol, in press.
4 Bangsbö J, Gollnick PD, Graham TE, Juel C, Kiens B, Mizuno M, Saltin B: Anaerobic energy production and O_2 deficit-debt relationships during exhaustive exercise in humans. J Physiol 1990;422:539–559.
5 Benade AJS, Heisler N: Comparison of efflux rates of hydrogen and lactate ions from isolated muscles in vitro. Resp Physiol 1978;32:369–380.
6 Bergström J, Hermansen L, Hultman E, Saltin B: Diet, muscle glycogen and physical performance. Acta Physiol Scand 1967;71:140–150.
7 Blomstrand E: Muscle metabolism during intensive exercise – influence of subnormal muscle temperature. Acta Physiol Scand 1985;125:suppl 547,
8 Bobbis LH: Metabolic aspects of fatigue during sprinting: in Macleod D, et al (eds): Exercise: Benefits, Limits and Adaptations. London, Spon, 1987; pp 116–140.
9 Bouhuys A, Pool J, Binkhorst RA, von Leeuwen P: Metabolic acidosis of exercise in healthy males. J Appl Physiol 1966;21:1040–1046.
10 Bouissou P, Defer G, Guezennec CY, Estrade PY, Serrurier B: Metabolic and blood catecholamine responses to exercise during alkalosis. Med Sci Sports Exerc 1988;20: 228–232.
11 Brien DM, McKenzie DC: The effect of induced alkalosis and acidosis on plasma lactate and work output in elite oarsmen. Eur J Appl Physiol 1989;58:797–802.

12 Bryant Chase P, Kushmerick MJ: Effects of pH on contraction of rabbit fast and slow skeletal muscle fibers. Biophys J 1988;53:935–946.
13 Cain DF, Davies RE: Breakdown of adenosine triphosphate during a single contraction of working muscle. Biochem Biophys Res Commun 1962;8:361–366.
14 Costill DL, Verstappen F, Kuipers H, Jansson E, Fink W: Acid-base balance during repeated bouts of exercise. Influence of HCO_3. Int J Sports Med 1984;5:228–231.
15 Danforth WH: Activation of the glycolytic pathway in muscle; in Chance B, Estabrook RW (eds): Control of Energy Metabolism. New York, Academic Press, 1965.
16 Dill DB, Edwards HT, Talbot JH: Alkalosis and capacity for work. J Biol Chem 1932;97:1.
17 Donaldson SKB, Hermansen L, Bolles L: Differential direct effects of H^+ on Ca^{2+} activated force of skinned fibres from the soleus, cardiac and adductor magnus muscles of rabbits. Pflugers Arch 1978;376:55–65.
18 Dudley GA, Terjung RL: Influence of acidosis on AMP deaminase activity in contracting fast-twitch muscle. Am J Physiol 1985;248:C43–C50.
19 Essen B: Glycogen depletion of different fibre types in human skeletal muscle during intermittent and continuous exercise. Acta Physiol Scand 1978;103:446–455.
20 Essen B, Henriksson J: Glycogen content of individual muscle fibres in man. Acta Physiol Scand 1974:90:645–647.
21 Fridén J, Seger J, Ekblom B: Topographical location of muscle glycogen: an ultrahistochemical study in the human vastus lateralis. Acta Physiol Scand 1989;135:381–391.
22 Jansson E, Dudley GA, Norman B, Tesch PA: ATP and IMP in single human muscle fibres after high intensity exercise. Clin Physiol 1987;7:337–345.
23 Gleeson M, Greenhaff PL, Maughan RJ: Influence of a 24 h fast on high intensity cycle exercise performance in man. Eur J Appl Physiol 1988;57:653–659.
24 Goldfinch J, McNaughton L, Davies P: Induced metabolic alkalosis and its effect on 400 m racing time. Eur J Appl Physiol 1988;57:45–48.
25 Greenhaff PL, Gleeson M, Maughan RJ: The effects of dietary manipulation on blood acid-base status and the performance of high intensity exercise. Eur J Appl Physiol 1987;56:331–337.
26 Greenhaff PL, Gleeson M, Whiting PH, Maughan RJ: Dietary composition and acid-base status: limiting factors in the performance of maximal exercise in man? Eur J Appl Physiol 1987;56:444–450.
27 Greenhaff PL, Gleeson M, Maughan RJ: Diet-induced metabolic acidosis and the performance of high intensity exercise in man. Eur J Appl Physiol 1988;57:583–590.
28 Greenhaff PL, Gleeson M, Maughan RJ: The effects of a glycogen-loading regimen on acid-base status and blood lactate concentrations before and after a fixed period of high intensity exercise in man. Eur J Appl Physiol 1988;57:254–259.
29 Greenhaff PL, Gleeson M, Maughan RJ: The effects of diet on muscle pH and metabolism during high intensity exercise. Eur J Appl Physiol 1988;57:531–539.
30 Greenhaff PL, Snow DH, Harris RC, Roberts CA: Bicarbonate loading in the thoroughbred horse: dose, method of administration and acid-base changes. Equine Vet J 1990;(suppl 9):83–85.

31 Greenhaff PL, Harris RC, Snow DH: The effect of sodium bicarbonate (NaHCO$_3$) administration upon exercise metabolism in the thoroughbred horse. J Physiol 1990; 420:69P.
32 Harris RC, Edwards RHT, Hultman E, Nordesjo L-O, Nylind B, Sahlin K: The time course of phosphorylcreatine resynthesis during recovery of the quadriceps muscle in man. Pflügers Arch 1976;367:137–142.
33 Harris RC, Sahlin K, Hultman E: Phosphagen and lactate contents of m. quadriceps femoris of man after exercise. J Appl Physiol 1977;43:852–857.
34 Harris RC, Marlin DJ, Snow DH: Metabolic response to maximal exercise of 800 and 2,000 m in the thoroughbred horse. J Appl Physiol 1987;63:12–19.
35 Heigenhauser GJF, Jones N: Bicarbonate loading; in Williams MH, Lamb DR (eds): Ergogenics: The Enhancement of Sports Performance. Carmel, Benchmark Press, 1991.
36 Hermansen L: Muscular fatigue during maximal exercise of short duration; in di Prampero PE, Poortmans J (eds): Physiological Chemistry of Exercise and Training. Basel, Karger, 1981, pp 45–52.
37 Hermansen L, Hultman E, Saltin B: Muscle glycogen during prolonged severe exercise. Acta Physiol Scand 1967;71:129–139.
38 Hirche HJ, Hombach V, Langohr HD, Wacker U, Busse J: Lactic acid permeation rate in working gastrocnemii of dogs during metabolic alkalosis and acidosis. Pflügers Arch 1975;356:209–222.
39 Hirvonen J, Rehunen S, Rusko H, Harkonen M: Breakdown of high energy phosphate compounds and lactate accumulation during short supramaximal exercise. Eur J Appl Physiol 1987;56:253–259.
40 Horswill CA, Costill DL, Fink WJ, Flynn MG, Kirwan JP, Mitchell JB, Houmard JA: Influence of sodium bicarbonate on sprint performance: relationship to dosage. Med Sci Sports Exerc 1988;20:566–569.
41 Hultman E: Studies on muscle metabolism of glycogen and active phosphate in man with special reference to exercise and diet. Scand J Clin Lab Invest 1967;19:suppl 94.
42 Hultman E, Bergstrom J, Anderson NM: Breakdown and resynthesis of phosphorylcreatine and adenosine triphosphate in connection with muscular work in man. Scand J Clin Lab Invest 1967;19:56–66.
43 Hultman E, Sahlin K: Acid-base balance during exercise; in Hutton RS, Miller D (eds): Philadelphia, Franklin Institute Press, 1980, vol 8, pp 41–128.
44 Hultman E, Sjoholm H: Energy metabolism and concentration force of human skeletal muscle in situ during electrical stimulation. J Physiol 1983;345:525–532.
45 Hultman E, Spriet LL, Sönderlund K: Energy metabolism and fatigue in working muscle: in Macleod D, et al (eds): Exercise: Benefits, Limits and Adaptations. London, Spon, 1987, pp 63–80.
46 Inbar O, Rotstein A, Jacobs I, Kaiser P, Dlin R, Dotan R: The effects of alkaline treatment on short-term maximal exercise. J Sports Sci 1983;1:95–104.
47 Jacobs I: Lactate, muscle glycogen and exercise performance in man. Acta Physiol Scand 1981 (suppl 495):1–35.
48 Jacobs MH: The production of intracellular acidity by neutral and alkaline solutions containing carbon dioxide. Am J Physiol 1920;53:457.

49 Jones NL, Sutton JR, Taylor R, Toews CJ: Effects of pH on cardiorespiratory and metabolic responses to exercise. J Appl Physiol 1977;43:959–964.
50 Kelman GR, Maughan RJ, Williams C: The effect of dietary modifications on blood lactate during exercise. J Physiol 1975;251:34–35.
51 Kelso TB, Hodgson DR, Witt EH, Bayly M, Grant BD, Gollnick PD: Bicarbonate administration and muscle metabolism during high intensity exercise; in Gillespie J, Robinson NE (eds): Equine Exercise Physiology. 2. Davis, ICEEP Publications, 1987, pp 438–447.
52 Kindermann W, Keul J, Huber G: Physical exercise after induced alkalosis (bicarbonate or Tris buffer). Eur J Appl Physiol 1977;37:197–204.
53 Klausen K, Sjogaard G: Glycogen stores and lactate accumulation in skeletal muscle of man during intense bicycle exercise. Scand J Sports Sci 1980;2:7–12.
54 Kowalchuk JM, Heigenhauser GJF, Jones NL: Effects of pH on metabolic and cardiorespiratory responses during exercise. J Appl Physiol 1984;57:1558–1563.
55 Kowalchuk JM, Maltais SA, Yamaji K, Hughson RL: The effect of citrate loading on exercise performance, acid-base balance and metabolism. Eur J Appl Physiol 1989;58:858–864.
56 McAuliffe JJ, Lind LJ, Leith DE, Fencl V: Hypoproteinemia alkalosis. Am J Med 1986;81:86–90.
57 McCance RA, Widdowson ED: The composition of foods. MRC Special Report Series No 297. London, HMSO, 1960.
58 McKenzie DC, Coutts KD, Stirling DR, Hoeben HH, Kuzara G: Maximal work production following two levels of artificially induced metabolic alkalosis. J Sports Sci 1986;4:35–38.
59 McNaughton L: Sodium citrate and anaerobic performance: implications of dosage. Eur J Appl Physiol 1990;61:392–397.
60 Mainwood GW, Worsley-Brown P: The effects of extracellular pH and buffer concentration on efflux of lactate from frog sartorius muscle. J Physiol 1975;250:1–22.
61 Maughan RJ: Effects of prior exercise on the performance of intense isometric exercise. Br J Sports Med 1988;22:12–15.
62 Maughan RJ, Poole DC: The effects of a glycogen-loading regimen on the capacity to perform anaerobic exercise. Eur J Appl Physiol 1981;46:211–219.
63 Maughan RJ, Leiper JB, Litchfield PE: The effects of induced acidosis and alkalosis on isometric endurance capacity in man; in Dotson Co, Humphrey JH (eds): Exercise Physiology. Current Selected Research. 2. New York, AMS Press, 1986, pp 73–82.
64 Nakamura Y, Schwartz A: Possible control of intracellular calcium metabolism by H^+: Sarcoplasmic reticulum of skeletal and cardiac muscle. Biochem Biophys Res Commun 1970;41:830–836.
65 Newsholme EA, Start C: Regulation in Metabolism. Chichester, Wiley, 1973.
66 Newsholme EA, Leech AR: Biochemistry for the Medical Sciences. Chichester, Wiley, 1983.
67 Noda L, Kuby SA, Lardy HA: Adenosine phosphate-creatine transphorphorylase. IV. Equilibrium studies. J Biol Chem 1954;210:83–85.
68 Osnes JB, Hermansen L: Acid-base balance after maximal exercise of short duration. J Appl Physiol 1972;32:59–63.

69 Parry-Billings M, Maclaren DPM: The effect of sodium bicarbonate and sodium citrate ingestion on anaerobic power during intermittent exercise. Eur J Appl Physiol 1986;55:524–529.
70 Poulus AJ, Doctor HJ, Westra HG: Acid-base balance and subjective feelings of fatigue during physical exercise. Eur J Appl Physiol 1974;33:207–213.
71 Ren J-M: Control of glycogenolysis, glucolysis, and contraction force in human skeletal muscle. Stockholm, Kongl Carolinska Medico Chirurgiska Institutet, 1989.
72 Richter EA, Galbo H: High glycogen levels enhance glycogen breakdown in isolated contracting skeletal muscle. J Appl Physiol 1986;61:827–831.
73 Robertson RJ, Falker JE, Drash AL, Swank AM, Metz KF, Spungen SA, LeBoeuf JR: Effect of blood pH on peripheral and central signals of perceived exertion. Med Sci Sports Exerc 1986;18:114–122.
74 Robin ED: Of men and mitochondria – intracellular and subcellular acid-base relations. N Engl J Med 1961;265:780.
75 Roedde S, MacDougall JD, Sutton JR, Green HJ: Supercompensation of muscle glycogen in trained and untrained subjects. Can J Appl Sport Sci 1986;11:42–46.
76 Rossing TH, Maffeo N, Fencl V: Acid-base effects of altering plasma protein concentration in human blood in vitro. J Appl Physiol 1986;61:2260–2265.
77 Sahlin K: Intracellular pH and energy metabolism in skeletal muscle in man: With special reference to exercise. Acta Physiol Scand 1978 (suppl 455):1–56.
78 Sahlin K: Effect of acidosis on energy metabolism and force generation in skeletal muscle; in Knuttgen HG, et al (eds): Biochemistry of Exercise. Champaign, Human Kinetics, 1983, pp 151–160.
79 Sahlin K, Alvestrand A, Brandt R, Hultman E: Acid-base balance in blood following exhaustive bicycle exercise and the following recovery period. Acta Physiol Scand 1978;104:370–372.
80 Sahlin K, Katz A: Purine nucleotide metabolism; in Poortmans JR (ed): Principles of Exercise Biochemistry. Basel, Karger, 1988, pp 120–139.
81 Saltin B, Gollnick PD: Skeletal muscle adaptability: significance for metabolism and performance; in Peachey LD (ed): Handbook of Physiology. Section 10. Baltimore, Waverley Press, pp 555–631.
82 Saltin B, Karlsson J: Muscle glycogen utilisation during work of different intensities; in Pernow B, Saltin B (eds): Muscle Metabolism during exercise. New York, Plenum Press, 1971, pp 289–300.
83 Siggaard-Andersen O: The Acid-Base Status of the Blood, ed 4. Copenhagen, Munksgaard, 1974.
84 Snow DH, Harris RC, Gash SP: Metabolic response of equine muscle to intermittent maximal exercise. J Appl Physiol 1985;58:1689–1697.
85 Spriet LL, Soderlund K, Bergstrom M, Hultman E: Skeletal muscle glycogenolysis, glycolysis and pH during electrical stimulation in men. J Appl Physiol 1987;62:616–621.
86 Spriet LL, Lindinger MI, McKelvie RS, Heigenhauser GJF, Jones NL: Muscle glycogenolysis and H^+ concentration during maximal intermittent cycling. J Appl Physiol 1989;66:8–13.
87 Stewart PA: How to Understand Acid-Base. New York, Elsevier, 1981, pp 1–186.
88 Stewart PA: Modern quantitative acid-base chemistry. Can J Physiol Pharmacol 1983;61:1444–1461.

89 Strome DR, Clancy RL, Gonzales NC: Contribution of a net transmembrane HCO_3^- flux to intracellular acid-base regulation. J Appl Physiol 1977;43:931–935.
90 Sutton JR, Jones NL, Toews CJ: Effect on pH on muscle glycolysis during exercise. Clin Sci 1981;61:331–338.
91 Swank A, Robertson RJ: Effect on induced alkalosis on perception of exertion during intermittent exercise. J Appl Physiol 1989;67:1862–1867.
92 Symons JD, Jacobs I: High-intensity exercise performance is not impaired by low intramuscular glycogen. Med Sci Sports Exerc 1989;21:550–557.
93 Wallace WM, Hastings AB: The distribution of the bicarbonate ion in mammalian muscle. J Biol Chem 1942;144:637.
94 Wilkes D, Gledhill N, Smyth R: Effect of acute induced metabolic alkalosis on 800-m racing time. Med Sci Sports Exerc 1983;15:277–280.

R.J. Maughan, MD, Department of Environmental and Occupational Medicine, University Medical School, GB-Aberdeen AB9 2ZD (UK)

Gastrointestinal Symptoms in Athletes: Physiological and Nutritional Aspects

Fred Brouns

Nutrition Research Center, Department of Human Biology, University of Limburg, Maastricht, The Netherlands

Introduction

During the last decade the participation in endurance events has become very popular. Large numbers of trained and untrained people participate in competitions, some of which last > 15 h. Physicians and first-aid teams frequently encounter participants suffering from abdominal pains, abnormal defecation, vomiting, nausea, etc. Although such problems seem to occur in some individuals in almost any endurance competition, it seems very difficult to indicate a specific cause because many factors play play a role in the etiology of a specific gastrointestinal (GI) symptom. The chance that a person suffering from a nonexercise-related GI disease such as peptic ulcer, gastritis, colonic spasms, etc., participates in a competition, increases with the number of participants in that competition. Such a person may experience abdominal pains during exercise, although exercise itself may not be the causative factor. Today more than 15,000 people may participate in major city marathons.

The chance that someone experiences an urge to defecate or has to urinate during exercise increases with increasing duration of the event. In this respect it is reasonable to assume that not wanting to stop running, in order to defecate or to urinate, will lead to abdominal distress and cramping, just as it will after some time in the resting state. However, there may be endurance athletes who experience GI symptoms for other, exercise- and diet-related, reasons. Highly trained marathon runners are known to prepare very carefully for each marathon. They have special training and diet regimens for the precompetition period and they pay particular attention to their food and fluid intakes on the day of the race. Nevertheless, experienced runners are still prone to GI distress as for example Derek Clayton who stated after setting his marathon world record in 1979: 'Two hours later the elation had warn off, I was urinating large dots of blood and I was vomiting black mucus and had a lot of black diarrhea. I don't think too many people can understand what I went through for the next 48 hours' [Runners World, May 1979, p 72]. Less well-trained, inexperienced runners may decide to imitate dietary precompetition regimens of top athletes and experience abdominal discomfort, simply because of the dramatic changes in quality and quantity of food intake compared to their normal daily diet. Recently a number of studies have reported on the frequency of GI symptoms in different types of endurance events and one has speculated on the possible causes of such symptoms. We will first give brief consideration to these studies.

GI Symptoms, Type and Frequency

In 1969, Dancaster et al. [24] described 2 well-trained distance runners who suffered from acute tubular necrosis and had diarrhea during and immediately after an endurance race. Fogoros [44] described 2 other cases of long-distance runners, suffering from loose bowel movements and bloody diarrhea. These 2 runners indicated that the symptoms occurred during periods of very intensive training or during all-out competitions. Cantwell [15] also reported 2 cases of bloody stools in runners. A 17-year-old runner experienced transient crampy abdominal pain, gas and light brown diarrhea after interval tempo training and bloody diarrhea 1 day later after 0.8- and 1.6-km tempo runs. A 27-year-old female runner experienced bloody diarrhea after a 16-km run. Sullivan [142] interviewed 57 long-distance runners. Thiry percent of these athletes indicated occasionally or frequently to suffer from an urge to defaecate while running, 25% suffered from abdominal cramps and/or diarrhea and 6% became nauseated. In 1984, Keeffe et al. [68] performed a more extensive study and carried out a questionnaire survey among 1,700 participants in a marathon race. A total of 707 runners (40%) responded.

The following results were obtained: (1) Lower GI symptoms were more commonly experienced with running than upper GI complaints. (2) The urge to defecate was the most common symptom experienced by runners (36.4% at moderate intensity (MI) to 38.6% at high intensity (HI)) and appeared both during and immediately after running. (3) Bowel movements (34.9%) and diarrhea (19.2%) were relatively frequent immediately after running. (4) Runners needed to interrupt runs for bowel movements, 16% (MI) and 18.4% (HI), and diarrhea, 8.2% (MI) to 10% (HI). (5) 1.2% (MI) and 2.4% (HI) of runners had bloody bowel movements. (6) Lower GI symptoms were more common in women than in men. (7) Some symptoms were more frequently reported by younger than older runners. (8) Nausea (11.6% (MI) and 12.7% (HI)) and vomiting (1.8%) occurred more during hard runs or after running.

In a more recent large survey, Riddoch and Trinick [112] received 471 completed questionnaires from a total study population of 1,750 competitors. Eighty-three percent of the respondents indicated to have suffered from one or more GI symptoms. Urge to defecate (53%) and diarrhea (38%) were the most common symptoms, especially among female runners (74 and 68%, respectively). Upper and lower GI symptoms were significantly more experienced by women than men and also more by younger than by older runners. Additionally, the participants reported that GI symptoms were more common during a 'hard' run than during an 'easy' run.

From the last two studies it becomes clear that the number of symptoms experienced among endurance athletes is considerable. However, the authors mention that primarily those athletes who suffer the most from GI problems may have responded to the questionnaire survey. In this case, the figure of 30–50% of runners who suffer from GI distress, as found in questionnaire studies, might be too high because of sample bias. This problem was taken into consideration by Rehrer et al. [110], who selected at random 44 subjects from a group of 114 previously untrained subjects at the beginning of an 18-month training regimen, aimed at completion of a 25-km run after 1 year of training and a marathon after 1½ years [64]. There were 12 women and 32 men in the group surveyed for GI symptoms, the ratio of women to men being similar to that of the total group. In the 25-km race it was found that the frequency of symptoms in women (58.3%) was significantly higher compared to men (12.5%). However, during the marathon the occurrence of symptoms was less different (56.3% in men, 41.7% in women). When analyzing the observed frequency of specific symptoms it was found that especially upper GI symptoms were experienced during the 25-km run. During the marathon, the frequency of upper GI symptoms increased further but lower GI tract-related symptoms also occurred (table 1). Further, it was observed that in 80% of all subjects

Table 1. Frequency of specific GI complaints among 44 runners surveyed (%) [data from 111]

	25 km	Marathon
Nausea	11	11
Stomachache	2	25
Abdominal stitches	16	23
Intestinal cramps	0	18
Defecation	0	2
Diarrhea	0	0
Urination	9	25

who were found to be dehydrated more than 4% suffered from GI symptoms. This symptom frequency was twice as high as in less dehydrated runners.

In two other marathon competitions (Belfast 1986 and 1987) Riddoch [113] selected 36 and 37 subjects prior to the race. Total symptom frequency was higher in 1986 (44%) compared to the 1987 marathon (19%). Symptoms described were stomach cramps, flatulence, diarrhea, heartburn, vomiting and nausea.

In another study [111], a survey was carried out among ultra-distance runners competing in the 67-km Swiss alpine marathon. One hundred and seventy participants agreed prior to the race to volunteer for a GI symptom study. Of these 170 subjects, the first 101 were recruited to give blood samples prior to the race and after the race. Complete blood samples before and after) were obtained from 89 subjects. Mean finishing times were 8 h 18 min for men and 8 h 56 min for the women. Forty-three percent of all subjects and 33% of those who gave the blood samples complained of GI distress during the race. Women experienced more lower GI symptoms than men (table 2).

The previously described studies have been done in endurance runners and the question arises whether runners experience more GI problems than other endurance athletes as for example cyclists. A group of athletes which are very useful in order to study this problem are triathletes.

Worobeth and Gerrard [161] performed a questionnaire study in 119 participants of an endurance event in which 800 m swimming, 25 km cycling, 5 km canoe and 12 km of running were performed. Fifty-nine percent of the subjects (n = 70) responded and the data showed that 58%

Table 2. Percentage runners (n = 170) participating in the 67-km Swiss alpine marathon, reporting GI problems [data from 111]

Symptoms	All	Male	Female
Nausea	15	14	25
Vomiting	2	2	0
Stomachache	11	12	0
Abdominal stitches	19	17	42
Intestinal cramps	9	8	25
Diarrhea	6	4	25*

* Significant difference between men and women ($p < 0.05$).

Table 3. Number of triathletes who experienced symptoms (n = 110) [data from 134]

Symptom	Portion of triathlon			Exercise interrupted because of symptoms
	swim	cycle	run	
Abdominal stitches	16	9	75	18
Gastroesophageal reflux	18	11	26	3
Nausea/vomiting	14	8	26	5
Abdominal cramps	15	7	38	9
Belching	36	12	32	1
Flatulence	12	14	52	–
Bowel dysfunction[1]	9	9	54	28
Bloody bowel movements	2	2	3	–
Urinary incontinence	5	4	19	2

[1] Diarrhea, urgent bowel movement, or faecal incontinence.

suffered from upper GI symptoms and 61% from more disturbing lower GI symptoms. These figures are about 20% higher than those of other studies. Again the possibility has to be considered that primarily the participants who suffered from GI symptoms may have responded as was the case in another study by Sullivan et al. [135], which surveyed 141 triathlon participants by a questionnaire. A total of 110 participants responded. These athletes appeared to have a low frequency of GI symptoms while swimming or cycling. Running, however, was associated with a high fre-

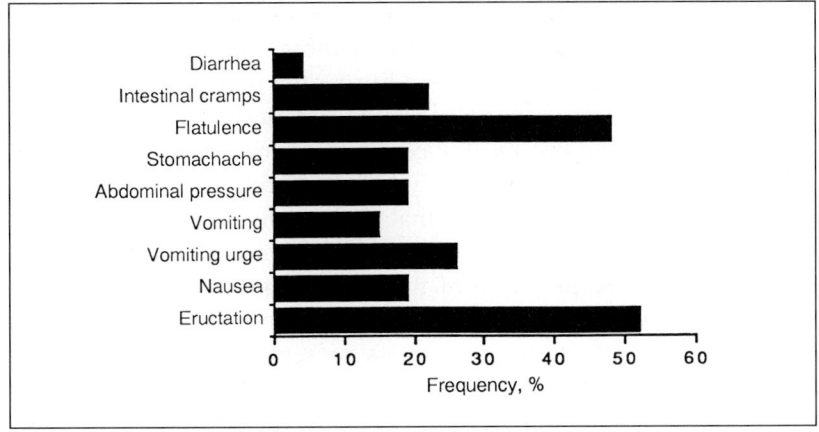

Fig. 1. Frequency (%) of GI symptoms in triathletes [data from 111].

quency of gastroesophageal and colonic symptoms. The need to defaecate was the most common reason for interrupting exercise (table 3). This indicates that the mode of exercise may be an important factor in the etiology of GI symptoms during exercise.

A detailed analysis of this data demonstrated that both upper and lower GI symptoms were affected by exercise intensity, recency of meals and intake of specific foods during exercise, such as high fiber foods, coffee and orange juice. Anxiety seemed to be an additional influence on the occurrence of lower GI symptoms. Further it was noted that women were more likely to experience abdominal cramps while exercising during their menses.

Recently, Worme et al. [160] analyzed GI symptoms in 21 female and 50 male triathletes. Upper GI symptoms occurred in 50% of all cases, whereas lower GI symptoms were described by only a few subjects. Twenty-seven percent of a subgroup of 30 athletes tested for fecal blood loss were positive. Aspirin, which may affect bleeding, was used by 45% of the total population. Rehrer et al. [110] tried to correlate the effect of dietary habits, including food and fluid consumption during triathlon competition, on the occurrence of GI symptoms. Forty-nine percent of the 55 triathletes studied had one or more GI symptoms, 29% one or more severe disturbing symptoms. The most common complaints were the less disturbing such as eructation and flatulence (fig. 1). In general, GI symptoms occurred more frequently in triathletes who ingested hypertonic carbohydrate solutions during exercise and in those who consumed foodstuffs high in dietary fiber, fat or protein during the precompetition meal. The last three factors were especially related to vomiting and reflux. In addition, the time of the last meal played a role. Triathletes who consumed a solid food-containing meal shortly before exercise suffered more vomiting. In general, it was observed that the complaints were more common during running and were least common during swimming. Care should be taken with the interpretation of this finding since the running part of a triathlon is at the end of the competition. In this respect, both time- and fatigue-related factors, which are less present during swimming at the onset of the competition, may be involved. One study is available in which 'a gliding type' of sport was performed. Kehl et al. [70] observed that 19.5% of 41 subjects participating in a cross-country ski marathon race suffered from abdominal pains and diarrhea during the competition.

In several studies the need to urinate has been considered as a symptom which interferes negatively with exercise. Many athletes rate the urge to urinate as a GI symptom. In our opinion this is not right. Urination is a normal physiological event, also when occurring during exercise. The chance of having to urinate increases with increasing exercise time and in addition with increasing fluid consumption. With respect to the latter, urination seems to be a common occurrence in slower runners who consume large amounts of fluid. However, it cannot be explained easily why athletes who run at high speeds and demonstrate decreased plasma volumes due to dehydration also experience urination. Although the urine volumes in these runners usually are very small, the urge to urinate may be very strong. A possible explanation in this case may be that the body attempts to eliminate the concentrated urine in order to eliminate urea and accumulated electrolytes. It seems logical that resistance to this physiological demand may cause upset of the lower abdominal region. In addition, there are a number of other factors which may initiate the occurrence of GI tract symptoms in the exercising individual. Sandell et al. [118] analyzed factors associated with collapse during and aftr ultramarathon foot races. The major factors observed were inadequate training, failing to ingest carbohydrate-rich diets prior to competition, hypoglycemia, not eating a breakfast on the race day, prerace illness and hypothermia. It is striking that all these factors are also known to be associated to the occurrence of GI symptoms.

Summarizing this first part, it can be concluded from the presently existing evidence that a number of different upper and lower GI symptoms occur in 30–50% of participants in endurance exercise, and that these symptoms may be related to more than one causal factor. These symptoms as well as possible causal factors are summarized below:

Upper GI symptoms
    Vomiting
    Reflux
    Bloating
    Stomach pain

Lower GI symptoms
    Intestinal cramping
    Urge to defaecate
    Diarrhea
    Bloody diarrhea
    Flatulence

Factors influencing the occurrence of GI symptoms
    Training status
    Exercise intensity
    Type of exercise
    Precompetition food intake
    Competition food/fluid intake
    Level of dehydration

Increased symptoms frequency observed
    In running activities
    At maximal exercise intensity
    In dehydrated runners
    After training interruption
    In younger athletes
    In females, especially when having menses

Two observations seem to be of particular interest, namely: (1) in sports events where the body is in a relatively stable position, such as during cycling, swimming, speed skating, or cross-country skiing, the frequency of abdominal complaints is lower than in running, *irrespective of nutritional intake before or during the event,* and (2) an improved training status seems to correlate with a decreased occurrence of GI disturbances.

It is impossible to completely elucidate the reasons of GI symptoms during exercise from the presently available scientific data. The underlying pathophysiology of GI disturbances associated with running has been little studied and remains speculative. This lack of knowledge, expressed by several authors, underlines the need for research [8, 44, 68, 80, 136, 140]. Nevertheless, when looking at the symptoms observed and the factors which may be of influence, it appears that changes in the following functions of the GI tract may be involved: (a) gastric emptying; (b) gut transit; (c) absorption in the gut, and (d) secretion in the gut. These functions are to a large extent regulated by nervous stimulation and hormonal processes which in turn are influenced by the immediate physiological activity level, i.e. metabolic demand and blood supply. The logical question then is: Does exercise alter these functions by influencing the regulating factors?

Exercise and GI Function

GI Blood Flow

When starting exercise there are a number of processes which induce physiological and biochemical adaptations to the higher degree of physical activity. Probably the most important alterations are immediate local changes in tissue biochemistry, such as changes in oxygen saturation, content of energy-rich phosphates and accumulation of metabolites, as well as changes in central regulation, such as a change in sympathetic and parasympathetic output and circulating hormones.

In general, these processes lead to vasodilatation in those tissues which undergo increased activity, require more oxygen, ATP resynthesis and elimination of metabolic end products, whereas vasoconstriction will appear in the relatively nonactive tissues. As a result, the muscles will be supplied with more blood and the GI tract will undergo a decreased blood supply during exercise [10, 19, 47, 71, 115, 151].

Rowell et al. [115] showed a 60–70% decrease in splanchnic blood flow in man, working at submaximal exercise intensity of 70% V_{O_2max}. Clausen [19] reported that during maximal exercise in both trained and untrained people, blood flow may be reduced to 20% compared to resting levels (100%). This quantitative decrease in blood flow seems to depend, to a large extent, upon the relative exercise intensity since there is a close relationship to the change in heart rate. In this respect, sympathetic output seems to play an important role in the redistribution of blood flow which occurs during exercise. As such, the fall in splanchnic blood flow at submaximal exercise intensities, as during endurance events, may be minimized by adequate training since this has been shown to reduce parasympathetic output [48]. This is further confirmed by the observation that training leads to a decreased heart rate for a given exercise intensity and that this occurs together with an increased indocyanine green clearance by the liver, indicative of an increased blood flow through the splanchnic circulation [20, 115]. Interestingly, GI symptoms occur less frequently after adequate training or when relative exercise intensity is reduced [42, 44]. On the other hand, blood flow may be decreased to critical levels by maximal sympathetic stimulation and maximal hormonal changes, which will take place during maximal exercise when hyperthermia, hypohydration, hypoglycemia, hypoxia or a combination of these factors may be present. The concentration of noradrenaline in plasma has been shown to vary inversely with the oxygen saturation of mixed venous blood. Hyperthermia and hypoglycemia lead to highly increased sympathetic stimulation and plasma catecholamine levels and thus may influence intestinal blood flow significantly [for review, see 48].

Additionally, dehydration may lead to a further shift in body fluid, in order to maintain a certain level of central blood supply. Ongoing dehydration may therefore lead to tissue dehydration and decreased capillary blood supply in the extremities, in peripheral tissues such as the skin, and may induce a drop in intestinal blood flow to practically zero. In addition, blood viscosity is known to be affected by exercise-induced hyperthermia. Plasma viscosity is temperature-dependent and shows a 10% increase at 40 °C. Whole blood viscosity, which may further affect gut capillary blood flow, is influenced by the status of hydration [148]. The interplay of a large number of factors determines actual splanchnic blood supply (fig. 2). This has been reviewed extensively by Granger et al. [53]. The question remains as to the importance of individual factors during endurance exercise. A reduction of blood supply below critical level will lead to an insufficient oxygen supply below critical level will lead to an insufficient oxygen supply on the one hand, so-called gut ischemia, and the accumulation of metabolic waste products on the other.

Qamar and Reed [108] observed a 43% decrease in mesenteric artery blood flow at the end of 15 min uphill walking exercise. The decrease was still 24% at 9 min and 9% at 30 min postexercise. When a liquid meal was ingested (390 kcal, 21.6 g protein, 15.4 g fat, 41.3 g carbohydrate) in the first 10 min of exercise, the meal-induced increase in blood flow was significantly smaller shortly after intake (5 min) but not later than when the meal was ingested without exercise. These findings suggest that the exer-

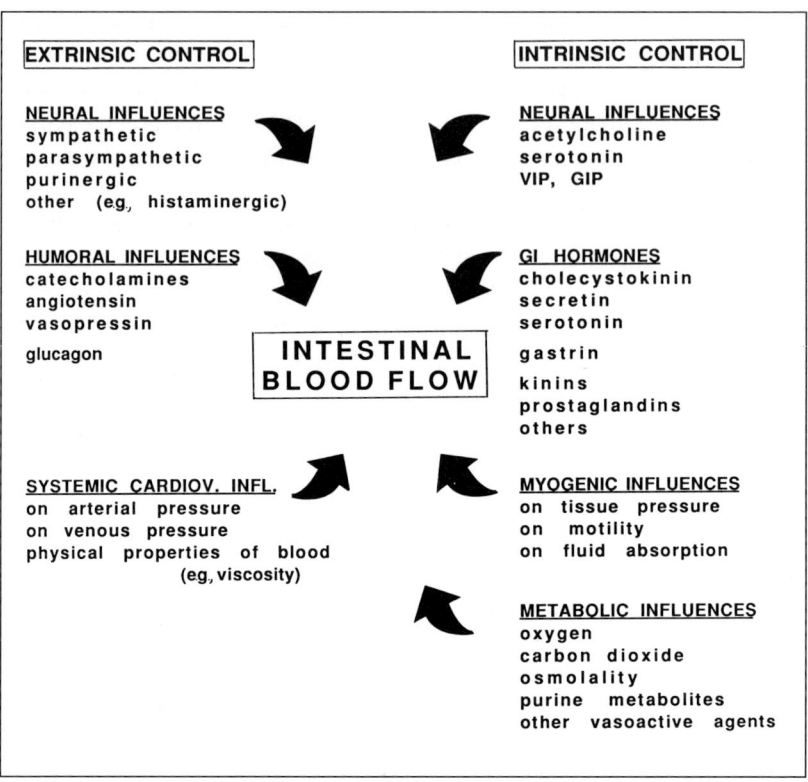

Fig. 2. Extrinsic control; major influences are: (1) neural, mainly sympathetic; (2) circulating vasoactive agents; (3) systemic hemodynamic changes. Intrinsic control; major influences are: (1) neural; (2) myogenic; (3) metabolic; (4) locally produced vasoactive substances [data from 53].

cise-induced reduction in mesenteric blood flow can be counteracted by meal-induced stimulation of blood flow, at least in a short-term submaximal exercise.

Blood Flow and GI Symptoms

It may be assumed that dehydration, reduced blood flow and hypoxia are of major influence in the occurrence of GI symptoms associated with the lower intestinal tract as frequently seen in ultra-endurance athletes. A lack of oxygen may induce acute energy deficits which may influence active, energy-dependent, absorption of nutrients, such as glucose or amino acids, and may also interfere with sodium-potassium pump capacity. As a result, absorption and cell integrity may become disturbed.

Studies on carbohydrate absorption in dogs showed that absorption was diminished when blood flow decreased [149, 157]. This may partly explain the observation that dehydrated runners react with GI upset immediately following consumption of concentrated carbohydrate solutions. A decreased absorption may lead to an increased osmotic load within the gut. As a result, gastric emptying may become further delayed by receptor feedback control [62, 63] and fluid secretion into the gut will take place to neutralize the difference in osmotic pressure. The latter may induce diarrhea. Recently, Rehrer [111] and Neufer et al. [94] showed that dehydration inhibits gastric emptying of carbohydrate-containing liquids.

Maughan et al. [81] observed delayed tracer accumulation in the blood, using deuterium oxide as tracer for absorption, when exercise was performed at 80% V_{O_2max}, compared to 42 or 61% V_{O_2max}. At all exercise intensities, tracer accumulation was less than at rest. It was speculated that strenuous exercise may reduce the availability of fluid ingested during exercise. In other studies [111, 152] it was observed that the amount of ingested carbohydrate which is oxidized is far less than the amount ingested and that the relative oxidation decreases with increasing carbohydrate load (fig. 3, 4). It is speculated that the nonoxidized fraction either is stored in unknown pools or remains in the gut. In the latter case osmotic fluid shifts and fermentation will take place which may induce diarrhea and flatulence. Interestingly these symptoms are most frequently reported after intake of more concentrated carbohydrate solutions in triathletes.

Gut ischemia may further lead to lesions of the inner surface of the gut and thus to intestinal bleeding and fluid secretion, as seen in cases of hypovolemic shock. There is some speculation about the site of this intestinal bleeding since experimental studies in animals show that an increased sympathetic activity leads to a closure of precapillary sphincters, reducing the actual flow capacity of the final capillary bed in the gut villi. As a result, metabolites accumulate in the capillary bed and start to exert a stimulating effect towards vasodilatation. As a result, the circulation through the villus tips will be 'reopened' at the cost of the blood flow through the deeper layers of the gut which remain vasoconstricted. This is called an autoregulatory escape to diminished blood flow through the villus tips, the absorptive site of the gut [53, 143]. However, since in absolute terms about 90% of total blood flow through the gut (in exercise this will be 90% from the

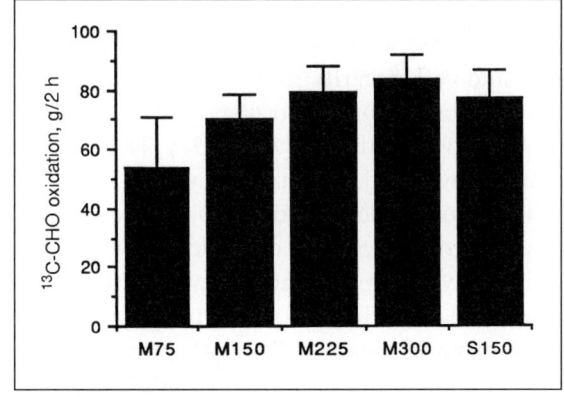

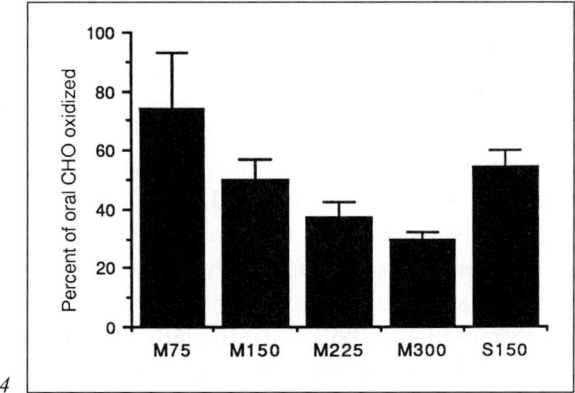

Fig. 3. Total carbohydrate oxidation after oral carbohydrate intake. M = Maltodextrin; S = sucrose. Figures indicate intake/g [data from 152].

Fig. 4. Percentage of total orally ingested carbohydrate. M = Maltodextrin; S = sucrose. The fraction which is oxidized decreases with increasing oral intake [data from 152].

remaining blood flow of 20%) is shunted to mucosa and submucosa [53], it may well be that the remaining blood supply to the villi is too small to maintain normal tissue integrity.

GI Bleeding

Some early cases of GI bleeding related to exercise have been reported. Fogoros [44] described 2 cases of runners with diarrhea, 1 of which was bloody. Both were associated with abdominal cramping and in both cases the runners had increased their training over a short period of time. Cantwell [15] reported a 17-year-old male runner who executed intensively anaerobic tempo training and experienced bloody diarrhea between the training session. Because of a tentative diagnosis of acute appendicitis the man was operated. The appendix was normal, but small bleeding points were observed on the serosa of the cecum and ascending colon. A second case was a 27-year-old female participating in a 16-km run; she passed bloody stools and suffered from crampy abdominal pain during the first 6 h after the run. Gut ischemia was proposed as possible cause. Thompson et al. [146] reported 1 case of exercise-related death, most probably caused by GI bleeding. Since then more cases have been reported and several studies have tried to explain causal factors. Porter [106] reported 3 cases of GI blood loss (Hemoccult-positive test) out of a group of 39 marathon participants. Keeffe et al. [68] reported visible blood loss in 1.8% of runners. A number of additional studies have shown that approximately 8–23% of runners have occult blood loss as shown by Hemoccult test [42, 54, 83, 84, 127]. Moses et al. [87] observed an extremely high percentage of ultra-distance runners (85%) with Hemoccult-positive postrace tests. Using the more specific Hemoquant test, Stewart et al. [139] observed 30% of runners to have fecal blood loss and 83% of runners to have increased fecal hemoglobin.

The use of the Hemoccult test, a qualitative stool test, has been criticized by Robertson et al. [114]. A comparison was made using three qualitative tests: Hemoccult, Hemacheck and Fecatwin S. The results of these tests were compared with the Hemoquant test, which gives quantitative figures. It was found that a large number of false-positive results were obtained with the Guaiacum card tests: Hemoccult (44%), Hemocheck (60%), Fecatwin S (57% false results), when compared with Hemoquant test.

Robertson et al. [114] studied subjects who were walking (4 times 37 km) or participating in a marathon. No GI bleeding occurred with walking. GI bleeding occurred in a significant number of marathon participants (mean faecal hemoglobin 0.42 mg/g in 28 subjects). Interesting in this study was that 13 marathon runners who used anti-inflammatory drugs prior to the start, in order to neutralize running-induced pain, had significantly more fecal hemoglobin (0.87 mg/g). The authors concluded, however, that although bleeding was significant, the quantitative loss was clinically unimportant. This was one of the first studies indicating that the use of inflammatory drugs can have adverse effects in runners because of its erosive effects on the gastric mucosa.

Schaub et al. [120] described a case of a 33-year-old marathon runner with complaints of nausea, urge to defecate and diarrhea (sometimes bloody) during HI training sessions or competition. Colonoscopic inspections showed epithelial surface lesions. These lesions were predominantly present in the cecum and colon. Because of the symptoms described and the colonoscopic findings, it was concluded that local gut hypoxia was the cause.

Moses et al. [87] also described ischemic colitis verified by colonoscopia. In this case the ischemic damage was restricted to the cecum and ascending colon. Schwartz et al. [125] endoscoped 3 out of 9 runners who were found to have a positive postexercise Hemoccult test. One subject had colonic hemorrhagic erosions and 2 others had hemorrhagic gastritis. A number of other reports have shown gastric hemorrhage as a cause of blood loss in runners [22, 57, 60, 100, 127, 129].

Sullivan [141] discussed a number of the factors which may be responsible for GI bleeding among which were mechanical factors, such a continuous mechanical jarring of the gut while running, so-called cecal slap syndrome [105] or a twist of the cecum due to the extreme low body fat of distance runners [107]. Additionally, ischemia was proposed not only to affect the gut but also the gastric wall. Psoas muscle hypertrophy, causing pressure on the ascending colon, when flexing the hip may be another rare cause of mechanical stress [26].

Since most studies have been performed on runners and running seems to be associated to a higher frequency of GI symptoms than 'gliding' sports such as cycling, skiing, marathon and speed skating, the question arises whether GI bleeding occurs less in gliding types of sport. Two studies are available. Kehl et al. [70] observed 3 subjects with fecal blood loss, out of 41 participants in a cross-country ski race. It was striking that these 3 subjects had trained less than 20 km/week, whereas all negatively diagnosed skiers had trained more (mean 38 km/week). It was observed that in 2 of the symptomatic subjects, blood flow through the mesenteric artery, measured by Doppler duplex scanning, was found to be dramatically increased after a 2 h exercise test. Dobbs et al. [33] followed 12 cyclists during a complete competitive season. Six cyclists occasionally had Hemoccult-positive stools which were always correlated with higher exercise intensities. Recently, GI bleeding has further been reviewed by Eichner [36].

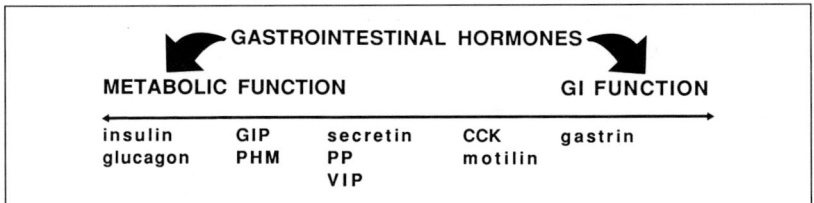

Fig. 5. Function of GEP hormones according to Buchanan [13]. GIP = Gastric inhibitory peptide; PHM = peptide-histidine-methionine, VIP = vasoactive intestinal peptide; CCK = cholecystokinin; PP = pancreatic polypeptide.

Gastroenteropancreatic (GEP) Hormones

Apart from the previously discussed catecholamines there are a number of so-called GEP hormones which are known to be elevated during exercise. Within the scope of this review, only some relevant aspects will be described. The reader can find more information in several extensive reviews [13, 17, 34, 117, 154, 158]. Although GEP hormones are mostly localized in the GI tract and pancreas, it does not necessarily mean that their only site of action is the GI tract. GI hormones are known to have both a GI function and/or a more general metabolic function (fig. 5). A part of the gut hormones are released by peptide-secreting nerve cells within the gut. Such 'hormones' in fact are neuropeptides [76]. Although many gut hormones are primarily stimulated or suppressed by food and fluid consumption, their metabolic properties lead us to the speculation that exercise can influence the secretion of GI hormones, irrespective of food consumption, because of exercise-induced demands with respect to energy metabolism and blood supply.

Several studies have tried to answer a possible effect of exercise on circulating GEP hormones. Hilsted et al. [61] studied 6 subjects during 3 h of low intensity exercise on a bicycle ergometer and observed significant increases in vasoactive intestinal peptide (VIP), pancreatic polypeptide (PP), somatostatin, glucagon and secretion. Sullivan [141] studied 7 well-trained runners during a 30-km run. He observed an increase in the same hormones, as well as in gastrin and motilin. Interesting in this study is the fact that VIP and glucagon increased continuously with exercise time. Oektedalen et al. [97] studied 12 participants in the ultra-distance WASA ski race (90 km) and observed significantly increased levels of VIP and secretin. More recently, two further studies have been carried out under com-

Table 4. Resting and postmarathon plasma levels (in ng·l^{-1}) of GI hormones in male (n = 32) and female (n = 7) marathon runners [data from 113]

Hormone	Control	Postmarathon
VIP (1986)	39±27	196±120***
VIP (1987)	31±14	152±90***
PHM	54±35	299±180***
NTW	5.3±3.8	3.5±2.5*
NTN	84±48	63±230**
Motilin	40±23	76±48***
Gastrin	45±24	134±84**
PP	160±124	494±260**
Secretin	15±9	46±15**

VIP = Vasoactive intestinal polypeptide; PHM = peptide-histidine-methionine; NTW = whole molecule neurotensin; NTN = N-terminal neurotensin; PP = pancreatic polypeptide. * $p < 0.05$; ** $p < 0.01$; *** $p < 0.001$.

Table 5. GI hormones (in ng·l^{-1}) measured in 89 participants of the 67-km Swiss alpine marathon [data from 110]

	Pre-race	Post-race
Gastrin	71.6±11.9	159.6±17.8*
VIP	28.8±2.6	224.3±20.1*
Motilin	146.4±13.0	214.1±15.1*
PHM	37.7±2.5	311.1±27.5*

* Significant postrace increase ($p < 0.05$).

petition circumstances. Riddoch [113] observed in participants of the Belfast marathon highly significant increases in VIP, peptide-histidine-methionine (PHM), gastrin and motilin (table 4). The same hormones were found to be highly increased in 89 subjects after running a 67-km Swiss alpine marathon [111] (table 5).

GEP Hormones and GI Symptoms

Some of the GEP hormones influence gut blood flow. VIP, glucagon and neurotensin increase, vasopressin decreases blood flow and thus may effect oxygen and energy supply. VIP, PHM, pancreatic glucagon and insulin influence lipolysis, glycogenolysis and gluconeogenesis. Based on these effects, these hormones may play an active role in metabolism during exercise. However, high levels of VIP and PHM are also known to produce watery diarrhea by reducing water and electrolyte absorption and enhancing water, chloride and bicarbonate secretion [67, 74, 75]. A blood plasma level of >150 ng·l^{-1} is seen as clinically diagnostic for patients with watery diarrhea hydrochloracidemia (WHDA syndrome) [82, 156], and the values observed in the ultra-distance studies are well beyond this level. It may thus be hypothesized that an increase of these peptides while exercising may cause lower GI symptoms in the athlete.

The values of the GEP hormones observed in the last two studies are much higher than reported before. In general, these data indicate that the magnitude of elevation of a number of GEP hormones may depend upon both duration and intensity of exercise and that the level of VIP and PHM can reach levels which in clinical situations are associated with the occurrence of diarrhea. Thus, symptoms described to occur as a result of heat exhaustion, such as vomiting, abdominal cramps and diarrhea, may be well explained as a result of the physiological changes which underlie heat exhaustion (hyperthermia, hypovolemia and induced hormonal changes), rather than heat exhaustion per se. The question remains as to why do these hormones increase so dramatically during exercise? Increased secretion, because of a metabolic stimulus, seems possible but is not proven. Another explanation was offered by Sullivan [141] who suggested that the reduction in splanchnic blood flow during exercise may lead to a decreased clearance of these hormones. Also a decreased inactivation by the kidneys may be of influence. Renal blood flow is inversely related to exercise intensity [50] and may at longlasting submaximal intensities be further affected by sweatloss-induced dehydration. Kidneys are the major site of inactivation of gastrin, GIP and PP [131]. Other explanations may be the mechanical stress caused by the continuous up and down movements of the gut while running but also intestinal ischemia. Interesting in this respect is that VIP and PHM mainly stem from neurons which are localized in the myenteric and submucosal plexuses of the gut. Mechanical stretching may lead to an increased release of VIP as was shown by Fahrenkrug et al. [39]. Riddoch [113] compared the effect of running and cycling in 9 male subjects, exercising at an intensity of 60% V_{O_2max}. He observed significantly increased levels of VIP and PP in running compared to cycling.

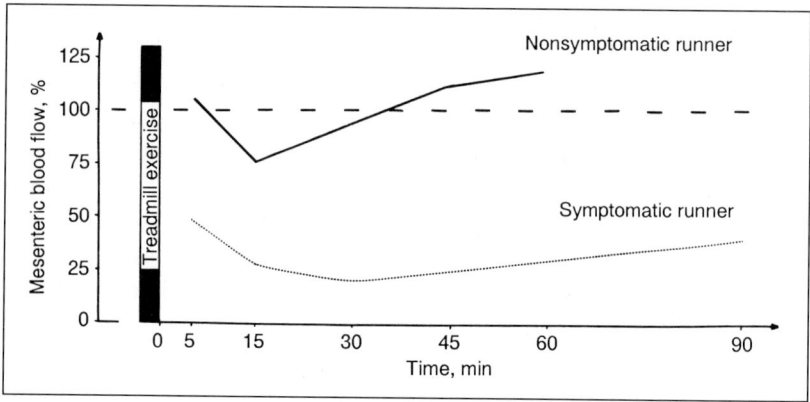

Fig. 6. Blood flow through the superior mesenteric artery after treadmill exercise in a symptomatic and a nonsymptomatic runner [data from 70].

It may further be that hypoxia stimulates neural activity and VIP release additionally. Hypoxia has been shown to reduce primarily blood flow in the deeper layers of the gut, which are the site of VIP-releasing neurons [132]. Modlin et al. [86] showed that experimentally induced ischemia induces a large VIP release, which may be an observation of crucial importance with respect to the interactions during exercise. Interesting in this respect is the study of Oektedalen et al. [96] who showed that increased VIP levels exist up to 2 h after the termination of exercise, although the half-life time of this peptide is assumed to be a few minutes only [35, 86]. If this finding is seen in the light of the observations done by Kehl et al. [70], who found that blood flow through the superior mesenteric artery in symptomatic runners may be decreased by as much as 80% up to more than 1 h postexercise (fig. 6), it may be assumed that continued postexercise local ischemia is responsible for the increased postexercise VIP levels observed.

Hormones and Gastric Secretion

Gastric acid reflux, indicated as heartburn, is a frequently occurring symptom in athletes suffering from GI distress. Anecdotally this symptom has often been suggested to be caused by an increased gastric acid production. However, evidence does not support this and does rather point to a change in motility and sphincter pressure as possible cause. The behavior

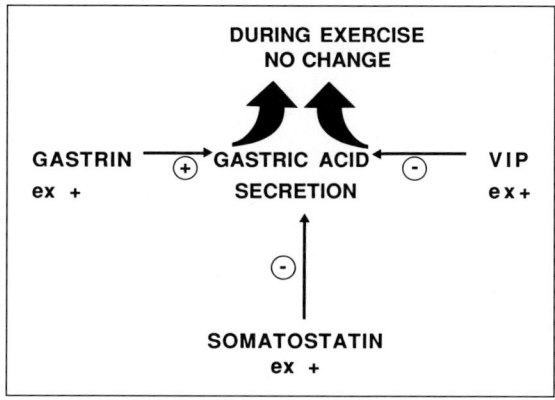

Fig. 7. Possible GEP hormone effect on gastric acid secretion during exercise.

of gastrin during exercise, a hormone which induces gastric acid secretion, is interesting in this respect.

Feldman and Nixon [41] observed that serum gastrin levels in response to a meal were significantly increased by exercise. They suggested that this may be explained by an exercise-induced catecholamine release, which induces gastrin secretion [18]. Also, Brandsborg et al. [11], Oektedalen et al. [98], Riddoch [113] and Rehrer [111] observed significant increases in serum gastrin in response to exercise. Surprising is however that this gastrin increase does not seem to lead to an increased gastric acid production in humans [41, 45, 79]. Additionally, exercise does not increase gastric pH [78, 79] or vagal tone, two other factors which would also normally stimulate acid secretion. In this respect it is hypothesized that the observed increase of both VIP and somatostatin, which are both known to inhibit gastric acid secretion, may overrule the effect of gastrin [113] (fig. 7).

So far, exercise-induced reflux cannot be explained by increased production of gastric acid. Thus, other factors must be responsible for reflux and 'heartburn' reported by endurance athletes. A recent study in humans showed an increased higher but decreased lower esophageal sphincter pressure as a result of exercise [28,103]. In another study [162], however, mean lower esophageal sphincter pressure was found to be increased. Studies on GI motility have shown that VIP induces a lowering of esophageal sphincter pressure in animals [109] and also inhibits antral motility [66], which is needed for gastric emptying. Clark et al. [21] and Kraus et al. [72, 73] measured esophageal pH during running, cycling and weight training and observed reflux induced by running and weight training, but not by cycling.

Further, it is observed that exertional gastroesophageal reflux with longlasting elevated postexercise esophageal pH occurred in a significant number (19%) of angina pectoris patients who had normal coronary angiograms. Another 25% had reflux associated with exercise with no postexercise elevation of pH. In general, reflux coincided with chest pain. Mean lower esophageal sphincter pressure was lower in both groups [122]. Additionally, exercise-induced reflux was observed in a child when performing intensive exercise [92]. Thus, a reduction in sphincter pressure and a reduced gastric emptying, as seen in highly intensive running exercise, but also in submaximal exercise in a dehydrated state [111], may well lead to reflux and heartburn. This is one of the indications that exercise may alter GI tone and motility.

Kidney Function

Although not linked directly to the GI region, the kidneys and the bladder may also be involved in lower abdominal discomfort. In the last decade a number of cases have been reported with acute renal failure associated to rhabdomyolysis (massive muscle tissue damage), all or not in a state of exertional heat stroke [1, 29, 30, 49, 51, 55, 77, 101, 121, 123, 124, 133, 150].

The factors supposed to be involved in the etiology of acute renal failure are extracellular volume reduction, hyperuricemia, hypokalemia, rhabdomyolysis with myoglobinuria and poor physical conditioning. The major cause of extracellular fluid depletion and hypovolemia in subjects performing endurance exercise in the heat is sweat loss, not compensated by adequate fluid consumption.

The latter indicates that exercise-induced fluid reduction not only threatens blood flow through the GI region, but also may damage the kidney. In the scope of this article, kidney failure will not be further discussed. The interested reader is referred to the references given or to the reviews of Anderson et al. [4]. Shibolet et al. [30] and Schrier et al. [123]. Blood loss via the urine may also originate from the bladder. Apart from renal and bladder trauma as reported in boxers, skiers and football players [46] another proposed mechanism [7] is mechanical jarring of the bladder due to the vertical shock moments associated with the landing phase of each stride. It is suggested that this may cause contusion, superficial bladder lesions and subsequent bleeding, especially when the bladder is empty [3, 7, 59], such as in a dehydrated state. Blacklock [7] proposes that this can be avoided by adequate filling of the bladder. A review of other mechanisms involved in changes in postexercise protein urea in humans is given by Poortmans [104].

Summarizing, it can be stated that exercise induces changes in GI blood flow which, when exercise is extremely intensive or when dehydration causes hypovolemia, may lead to local hypoxia of the GI tract and increased neural activity of the plexus submucosa of the gut. As a result, the secretion of some of the GEP hormones may be enhanced and decreased absorption parallel with increased secretion may take place. Diarrhea, intestinal cramping, delayed gastric emptying and bleeding of the serosa of the stomach and colon may result. This bleeding may be aggravated by the use of anti-inflammatory drugs.

GI Motility

Reflux, vomiting, diarrhea and cramps are clear indications of changes in GI motility and transit. Does exercise alter motility directly?

Gastric Emptying

During the last decade, a substantial number of studies have been done with respect to the effect of exercise on gastric emptying of liquids and half solids. These studies have partly been reviewed [12, 93]. More recently, a number of studies have been performed using a double sample technique which allows for frequent sampling and determination of gastric content and gastric secretion [6]. Results of these studies will be presented in Chapter 9 of this volume. For that reason, only the most important points will be summarized here.

In general, the existing evidence shows that exercise with intensities up to about 70% V_{O_2max} do not influence gastric emptying. Carbohydrate concentration and thus energy density plays a role, more concentrated solutions leaving the stomach at a slower rate than less concentrated solutions. In general, it is found that solutions containing up to approximately 80 g carbohydrate/l empty during exercise at about an equal rate as plain water. The addtion of carbohydrate and sodium to a fluid will stimulate the absorption of water in the gut and will also contribute to energy metabolism. There seems to be no real effect of training status or type of exercise, although running may lead to a decreased gastric emptying rate at a later stage [111]. Exercise at intensities higher than 70% V_{O_2max} tends to delay gastric emptying. The latter findings may be explained by the factors described in the previous two sections.

Apart from physical stress leading to the secretion of catecholamines and endogenous endorphins, emotional strain may also have a strong influence on gastric emptying rate. Beaumont [5] was probably one of the first to report that emotions can delay gastric emptying. In several recent studies [137, 138, 145, 146] it was shown that cold pain and labyrinth stimulation caused a marked delay in gastric emptying of a liquid meal and also altered the secretory response to that meal. The stress led to an increase in noradrenaline and β-endorphins. The authors speculate that these changes may be part of a systemic response to stress. Psychological stress [25, 32, 159] and exhausting endurance exercise [2, 9, 56a] also increase these humoral acting substances. Catecholamines inhibit gut motility and endorphins may interact with opiate receptors in the gut [102], decreasing propulsive contractions and giving rise to spasms which may delay gastric emptying [16, 27, 142]. Additionally, their role in the regulation of basal absorption secretion is also under discussion. It is believed that endorphins can reduce water and electrolyte absorption in small intestine and colon [43]. Acoustic stress is also found to delay gastric emptying and to increase the release of gastrin, pancreatic polypeptide and somatostatin. It is speculated that the effect on the GEP hormones might be a direct stress effect [52]. In this respect, one can only speculate about how much emotional or physical stress, prior to the start or during exercise when coping with exhaustion, may affect the GI motility, secretion and absorption. Also prostaglandins, which may be elevated during exercise, may affect motility and are thought to be diarrheogenic [40, 116, 119]. It needs to be further elucidated to what extent exercise affects the release of these substances in quantities sufficient to exert functional affects on the gut. Recent extensive reviews with respect to GI function and regulation can be found in Johnson et al. [65].

Gut Transit

The question arises whether symptoms such as urge to defecate, defecation and loose stools can be explained directly by altered motility and gut transit. A change in motility of stomach or gut may affect gastric emptying and propulsive actions of the gut moving its contents from duodenum to rectum. Such effects could result from acute exercise. Weisbrot et al. [155] observed that gastric emptying is partly regulated by the contractile force of the antrum and the duodenum. A greater rate of emptying occurs when there is relatively high antral but low duodenal activity. Slow emptying occurred with high duodenal but low antral activity. Long-term effects have been suggested to be induced by a period of training.

Acute Exercise

Studies showing an acute effect of exercise on lower esophageal sphincter pressure have been described before. Cammack et al. [14] were among the first to study GI transit time in exercise. They used the breath hydrogen technique, which gives an indication of how soon the 'head' of a meal reaches the colon and becomes part of fermental degradation. They studied 7 subjects performing a 60-km ride which was intermittently divided over a 6 h period. No effect on transit was found. Exercise intensity however was low (heart frequency 117/min, pedaling rate 33/min) equivalent to approximately 30% V_{O_2max}. Keeling and Martin [69] used liquid meals during walking exercise in 12 male subjects (5.6 km/h, up a 2% grade) and observed a significant decrease in transit time using the same technique.

Transit through the gut is stimulated by contraction waves which longitudinally pass the gut. The occurrence of these so-called migrating motor complex (MMC) contractions was studied by Evans et al. [37], using radiotelemetry capsules. The authors observed a decreased occurrence of MMC, which would indicate a decreased transit activity. Ollerenshaw et al. [99] studied the effect of three different exercise intensity levels on mean intestinal transit time using labelled resin beads. No differences in transit time were observed. Meshkinpour et al. [85], also using a breath hydrogen technique, compared the effect of walking exercise with sitting at rest and found transit time to be increased from 55 min (at rest) to 89 min. The latter study is criticized by Moses [91], because a substantial number of their subjects were females and no mention of menstrual status (which is known to affect transit time by delayment in the luteal phase [153]) was made. However, Moses et al. [89] obtained similar results in a study of 10 volunteers who performed a 2 h long treadmill run at 65% V_{O_2max} and ingested water or a glucose polymer solution. Running delayed transit of both meals.

Additionally two studies have been carried out to determine the effect of exercise on small bowel motility. In 1980, Harrison [56b] studied transit time of an oral dose of radiopaque markers in 11 subjects who were physically active (jogging) for 25–45 min/day. No consistent differences were found between the control period and activity period. Evans et al. [37] studied the transit of telemetry capsules during submaximal exercise. Transit time was reduced.

Although the results from intestinal transit studies are conflicting, the majority of studies, however, indicate an exercise-induced decrease in gut transit time, which does not explain the occurrence of loose stools or urge to defecate. A possible explanation for the urge to defecate may be derived from the early work of DeYoung et al. [31], who studied colonic tone in exercising dogs, using a balloon technique. It was observed that colonic tone as well as motility rose shortly after exercise. This was most usually accompanied by defecation. Further, PHM and VIP may decrease internal anal sphincter pressure [95] and an exercise-induced decrease in insulin and an increase in pancreatic polypeptide are expected to inhibit colonic motility [144]. However, Cordain et al. [23] hypothesized that during exercise sympathetic stimulation may cause a relaxation of the GI tract which may facilitate the passage of contents from the colon into the rectum, due to the up-and-down bouncing motion associated with running, causing the need to defecate.

Effect of Training

Cordain et al. [23] looked at the effect of a 6-week aerobic running program (3 times 30 min at 70–80% V_{O_2max}/week) on bowel transit using carmine dye capsules and observed a 22.8% decrease in transit time.

Summarizing the above, it may be postulated that exercise reduces GI tone and motility, and tends to delay transit. A reduction in tone may lead to reflux from the stomach or the loose stools escaping from the colon. Gastric emptying is not influenced at moderate exercise intensities, but will be inhibited at maximal exercise intensity or in a state of dehydration. The latter probably induces a cascade of changes in hormone secretion and nervous output, leading to a change in absorption, secretion and gut motility. Additionally, severe emotional/mental strain may further influence GI function.

Conclusions and Prevention of GI Symptoms

Conclusions with respect to the occurrence, frequency, type of symptom and possible related causes have been summarized at the end of the specific paragraphs.

The following measures can be taken to reduce the occurrence of GI distress: (1) Adequate conditioning – appropriate increase in training volumes and intensity. (2) Training of drinking regimens during exercise. (3) Do not ingest strongly hypertonic carbohydrate solutions during exercise (> 800 mosm). (4) Ingest a carbohydrate-rich diet, but avoid dramatic precompetition changes in the diet. (5) Do not ingest fiber-rich foods prior to competition. (6) Limit fat and protein intake in the precompetition meal. (7) Do not ingest extra caffeine, high doses of vitamin C, bicarbonate or dubious nutritional ergogenic aids just prior to or during exercise. (8) Defecate and urinate shortly prior to exercise. (9) Do not use anti-inflammatory drugs to prevent exercise-induced pains. (10) Do not use a GI medication without prescription from your sports medical advisor. (11) Go for a gastroenterological examination when suffering recurrently from GI symptoms also occurring in the nonexercising state.

Acknowledgements

I gratefully thank Prof. C. Williams and Dr. C. Riddoch for their comments and suggestions in reviewing this chapter and Ms. Lilo Kauer for her help and expertise in preparing the manuscript.

References

1 Albertazzi A, Del Rosso G, Cappelli P: Acute renal failure after repeated physical stress. Lancet 1984;ii:1418–1419.
2 Allen M: Activity-generated endorphins: a review of their role in sports science. Can J Appl Sport Sci 1983;8:115–133.
3 Alvarez C, Mir J, Obaya S, Fragoso M: Hematuria and microalbuminuria after a 100-kilometer race. Am J Sports Med 1987;15:609–611.
4 Anderson RJ, Reed G, Knockel J: Heatstroke. Adv Intern Med 1983;28:115–140.
5 Beaumont W: Experiments and observations on the gastric juice and the physiology of digestion (1938); reprinted by Andrew Combe, Edinburgh 1983.
6 Beckers EJ, Rehrer NJ, Brouns F, et al: Determination of total gastric volume, secretion and residual meal using the double sampling technique of George. Gut 1988;29:1725–1729.
7 Blacklock NJ: Bladder trauma in the long distance runner; 10,000 meters haematuria. Br J Urol 1977;49:129–132.
8 Bortoff A: Influence of exercise on gastrointestinal function; in Horton ES, Terjung RL (eds): Energy, Nutrition and Energy Metabolism. New York, Macmillan, 1988, pp 159–171.

9 Bortz WM, Angwin P, Mefford IN, Boarder MR, Noyce N, Barchas JD: Catecholamines, dopamine and endorphin levels during extreme exercise. N Engl J Med 1981;305:466–467.
10 Bradley SE: Variations in hepatic blood flow in man during health and disease. N Engl J Med 1949;240:456–461.
11 Brandsborg O, Christensen NJ, Galbo H, et al: The effect of exercise, smoking and propanolol on serum gastrin in patients with duodenal ulcer and in vagotomized subjects. Scand J Clin Lab Invest 1978;38:441–446.
12 Brouns F, Saris WHM, Rehrer NJ: Abdominal complaints and gastrointestinal function during long-lasting exercise. Int J Sports Med 1987;8:175–189.
13 Buchanan KD: Gastrointestinal hormones: general concepts. Clin Endocrinol Metab 1979;8:249–263.
14 Cammack J, Read NW, Cann PA, Greenwood B, Holgate AM: Effect of prolonged exercise on the passage of a solid meal through the stomach and small intestine. Gut 1982;23:957–961.
15 Cantwell JD: Gastrointestinal disorders in runners. JAMA 1981;281:1404–1405.
16 Chapman WP, Rowlands EN, Jones CM: Multiple balloon kymographic recording of the comparative action of demerol, morphine and placebos on the motility of the upper small intestine in man. N Engl J Med 1950;243:171–177.
17 Chey WY, Gutierrez JG: The endocrine control of gastrointestinal function. Adv Intern Med 1978;23:61–84.
18 Christensen K, Stadil F: Effect of epinephrine and norepinephrine on gastric release and gastric secretion of acid in man. Scand J Gastroenterol 1976;37:441–446.
19 Clausen JP: Effect of physical training on cardiovascular adjustments to exercise in man. Physiol Rev 1977;57:779–815.
20 Clausen JP, Klausen K, Rasmussen B, Trap-Jensen J: Central and peripheral circulatory changes after training of the arms or legs. Am J Physiol 1973;225:675–682.
21 Clark CS, Kraus D, Sinclair J, Castell D: Vigorous exercise induces gastroesophageal reflux. Gastroenterology 1988;94:A612.
22 Cooper DT, Douglas SA, Firth LA, Hannagan JA, Chadwick VS: Erosive gastritis and gastrointestinal bleeding in a female runner: prevention of bleeding and healing of the gastritis with H_2-receptor antagonist. Gastroenterology 1987;92:2019–2023.
23 Cordain L, Latin RW, Behnke JJ: The effects of an aerobic running programme on bowel transit time. J Sports Med 1986;26:101–104.
24 Dancaster CP, Duckworth WC, Roper CJ: Nephropathy in marathon runners. South Afr Med J 1969;43:758–760.
25 Danner SA, Endert E, Koster RW, Dunning AJ: Biochemical and circulatory parameters during purely mental stress. Acta Med Scand 1981;209:305–308.
26 Dawson DJ, Kahn AN, Shreeve DR: Psoas muscle hypertrophy: mechanical cause for 'jogger's trots'? Br Med J 1985;291:787–788.
27 Davis TP, Culling AJ, Schoemaker J, Galligan JJ: Beta-endorphin and its metabolites stimulate motility of the dog small intestine. J Pharmacol Exp Ther 1983;227:499–507.
28 Demeirleir K, Peeters P, Peeters O, Cherys O, Clarys J, Smekens L, Devis G: Esophageal function during dynamic exercise. Med Sci Sports Exerc 1990;22:abstr S95.
29 Demos MA, Gitin EL: Acute exertional rhabdomyolysis. Arch Intern Med 1974;133:233–239.

30 Demos MA, Kitin EL, Kagen LJ: Exercise myoglobinemia and exertional rhabdomyolysis. Arch Intern Med 1974;134:669–673.
31 DeYoung VR, Rice HA, Steinhaus AH: Studies in the physiology of exercise. VII. The modificaiton of colonic motility induced by exercise and some indications for a nervous mechanism. Am J Physiol 1931;99:52–63.
32 Dimsdale JE, Moss J: Short-term catecholamine response to psychological stress. Psychosom Med 1980;42:493–497.
33 Dobbs TW, Akings M, Ratliff R, Eichner ER: Gastrointestinal bleeding in competitive cyclists. Am J Sports Med 1988;20:S78.
34 Dockray GJ: Physiology of enteric neuropeptides; in Johnson LR, Christensen J, Jackson MJ, Jacobson ED, Walsh JH (eds): Physiology of the Gastrointestinal Tract. New York, Raven Press, 1987, vol 1, pp 41–66.
35 Domschke WLG, Domschke S, Struntz U, Bloom SR, Walsch E: Effects of vasoactive intestinal peptide on resting and pentagastrin stimulated lower esophageal sphincter pressure. Gastroenterology 1978;75:9–12.
36 Eichner ER: Gastrointestinal bleeding in athletes. Physician Sports Med 1989;17:128–140.
37 Evans DF, Foster GE, Hardcastle JD: Does exercise affect small bowel motility in man. Gut 1989;24:A1012.
38 Evans DF, Foster GE, Hardcastle JD: Does exercise affect the migrating motor complex in man? In Roman C (ed): Gastrointestinal Motility. Boston, MTP Press, 1984, pp 277–284.
39 Fahrenkrug J, Haglund U, Jodal M, Lundgren O, Olbe L, Schaffalitzky de Muckadell OB: Nervous release of vasoactive intestinal polypeptide in the gastrointestinal tract of cats: Possible physiological implications. J Physiol (Lond) 1978;284:291–305.
40 Fargeas MJ, Fioramonti J, Bueno L: Prostaglandin E_2: a neuromodulator in the central control of gastrointestinal motility and feeding behavior by calcitonin. Science 1984;225:1050–1051.
41 Feldman M, Nixon JR: Effect of exercise on postprandial gastric secretion and emptying in humans. J Appl Physiol 1982;53:851–854.
42 Fisher RL: Exercising the gut – Therapy or complications? Editorial. Am J Gastroenterol 1986;81(April):299–300.
43 Fogel R, Kaplan RB: Role of enkephalins in regulation of basal intestinal water and ion absorption in the rat. Am J Physiol 1984;246:G386–G392.
44 Fogoros RN: Gastrointestinal disturbances in runners. 'Runners trots.' JAMA 1980;243:1743–1744.
45 Fordtran JS, Saltin B: Gastric emptying and intestinal absorption during prolonged severe exercise. J Appl Physiol 1967;23:331–335.
46 Frey U: The relationship between anatomical site of injury and particular sports. Proc R Soc Med 1969;62:917–919.
47 Fronek K: Combined effect of exercise and digestion on hemodynamics in conscious dogs. Am J Physiol 1970;218:555–559.
48 Galbo H (ed): Gastroenteropancreatic hormones; in Hormonal and Metabolic Adaptation to Exercise. New York, Thieme, 1983, pp 59–61.
49 Goldsmith HJ: Acute renal failure after a marathon run. Lancet 1984;i:278–279.
50 Grimby: J Appl Physiol 1965;30:1294–1298.

51   Grossman RA, Hamilton RW, Morse BM, et al: Non-traumatic rhabdomyolysis and acute renal failure. N Engl J Med 1974;291:807–811.
52   Gué M, Peeters T, Depoortere I, et al: Stress-induced changes in gastric emptying, postprandial motility and plasma gut hormone levels in dogs. Gastroenterology 1989;97:1101–1107.
53   Granger DN, Richardson PDI, Kvietys PR, Mortilaro NA: Intestinal blood flow. Gastroenterology 1980;78:837–863.
54   Halvorsen FA, Lyng J, Ritland S: Gastrointestinal bleeding in marathon runners. Scand J Gastroenterol 1986;21:493–497.
55   Hamilton RW, Garnder LB, Pen AS, et al: Acute tubular necrosis caused by exercise-induced myoglobulinuria. Ann Intern Med 1972;77:77–82.
56a   Harber VJ, Sutton JR: Endorphins and exercise. Sports Med 1984;1:154–174.
56b   Harrison RJ, Leeds AR, Bolster NR, Judd PA: Exercise and wheat bran; effect on whole gut transit. Proc Nutr Soc 1980;39:22A.
57   Heer M, Repond F, Hany A, Sulser H: Hemorrhagic colitis, gastritis, hematuria and rhabdomyolysis: manifestations of multisystemic ischaemia? Schweiz Rundsch Med Prax 1986;75:1538–1540.
58   Heer M, Repond F, Hany A, Sulser H, Kehl O, et al: Acute ischemic colitis in a female long-distance runner. Gut 1987;28:896–969.
59   Herbert LF, Natelson EA: Grossly bloody urine of runners. South Med J 1977;70: 1394–1396.
60   Hilpert G, Gaudin B, Devars Du Mayne JF, Cerf M: Gastrite ulcéreuse chez un coureur de fond. Gastroentérol Clin Biol 1984;8:983.
61   Hilsted J, Galbo H, Sonne B, Schwartz T, Fahrenkrug J, Schaffalitzky de Muckadell OB, Lauristen KB, Tronier B: Gastroenteropancreatic hormonal changes during exercise. Am J Physiol 1980;239:G136–G140.
62   Hunt JN, Pathak JO: The osmotic effects of some simple molecules and ions on gastric emptying. J Physiol 1960;154:254–269.
63   Hunt JN: The site of receptors slowing gastric emptying in response to starch in test meals. J Physiol 1960;154:270–276.
64   Janssen GME, Graef CJJ, Saris WHM: Food intake and body composition in novice athletes during a period to run a marathon. Int J Sports Med 1989;10(suppl 1): S17–S22.
65   Johnson LR, Christensen J, Jackson MJ, Jacobson ED, Walsh JH: Physiology of the Gastrointestinal Tract. New York, Raven Press, 1987, vol 1, 2.
66   Kachelhoffer J, Mendel C, Dauchel J, Hohmatter D, Grenier JF: The effects of VIP on intestinal motility: Study on ex vivo perfused isolated canine jejunal loops. Am J Dig Dis 1976;21:957–962.
67   Kane MG, O'Dorisio TM, Krejs GJ: Production of secretory diarrhea by intravenous infusion of vasoactive intestinal polypeptide. N Engl J Med 1983;309:1482–1485.
68   Keeffe EB, Lowe DK, Gross JR, et al: Gastrointestinal symptoms of marathon runners. West J Med 1984;141:481–484.
69   Keeling WF, Martin BJ: Gastrointestinal transit during mild exercise. J Appl Physiol 1987;63:978–981.
70   Kehl O, Jäger K, Münch R, et al: Mesenteriale Ischämie als Ursache der Jogging-Anämie. Schweiz Med Wochenschr 1986;116:974–976.

71 Konturek S, Falser J, Obtulowicz W: Effect of exercise on gastrointestinal secretions. J Appl Physiol 1973;34:324–328.
72 Kraus B, Sinclair J, Castell D: Distance running induces gastroesophageal reflux. Gastroenterology 1989;96:A685.
73 Kraus B, Sinclair J, Castell D: Ranitidine reduces distal esophageal acid exposure in runners. Gastroenterology 1989;96:A686.
74 Krejs GJ, Fordtran J: Effect of VIP infusion on water and ion transport in the human jejunum. Gastroenterology 1981;78:722–727.
75 Lancet: VIP and diarrhea. Lancet 1984;i:202.
76 Larsson L, Fahrenkrug J, Schaffalitzky de Muckadell OB, Sundler F, Hakanson R, Rehfeld JF: Localization of vasoactive intestinal polypeptide to central and peripheral neurons. Proc Natl Acad Sci USA 1976;73:3197–3200.
77 Lonka L, Pedersen RS: Fatal rhabdomyolysis in marathon runners. Lancet 1987;i:857.
78 Lorber SH: Gastrointestinal disorders and exercise; in Bone AA, Lowenthal DT (eds): Exercise Medicine: Physiological Principles and Clinical Application. Academic Press, New York, 1983, pp 279–290.
79 Markiewicz K, Cholewa M, Lukin M: Gastric basal secretion during exercise and restitution in patients with chronic duodenal ulcer. Acta Hepatogastroenterol 1979;26:160–165.
80 Maron MB, Horvath SM: The marathon: a history and review of the literature. Med Sci Sports Exerc 1978;10:137–150.
81 Maughan RJ, Leiper JB, McGaw A: Effects of exercise intensity on absorption of ingested fluids in man. Exp Physiol 1990;75:419–421.
82 Marx M, Newman JB, Guice KS, Nealon WH, Townsend CM Jr, Thompson JC: Clinical significance of gastrointestinal hormones; in Thomson JC (ed): Gastrointestinal Endocrinology. New York, McGraw-Hill, 1987.
83 McMahon LF, Ryan MJ, Lareen D, Fisher RL: Occult gastrointestinal blood loss in marathon runners. Ann Intern Med 1984;100:836–837.
84 McCabe ME, Peura DA, Kadakia SC, Bocek Z, Johnson LF: Gastrointestinal blood loss associated with running a marathon. Dig Dis Sci 1986;31:1229–1232.
85 Meshkinpour H, Kemp C, Fairster R: The effect of aerobic exercise on mouth to cecum transit time. Gastroneterology 1989;96:938–941.
86 Modlin IM, Bloom SR, Mitchell S: Plasma vasoactive intestinal polypeptide levels and intestinal ischaemia. Experientia 1978;34:535–536.
87 Moses FM, Baska R, Graeber G, Kerany PD: Gastrointestinal bleeding during an ultramarathon. Med Sci Sports Exerc 1989;21:578.
88 Moses FM, Brewer TG, Peura DA: Running-associated proximal hemorrhagic colitis. Ann Intern Med 1988;108:385–386.
89 Moses FM, Ryan C, DeBolt J, Smoak B, Hoffman A, et al: Oral-cecal transit time during a 2-hour run with ingestion of water of glucose polymer. Am J Gastroenterol 1988b;83:1055.
90 Moses FM, Singh A, Villanueva V, Kelsey B, Smoak B, et al: Lactose absorption and transit during prolonged high intensity running. Am J Gastroenterol 1989;84:1192.
91 Moses FM: The effect of exercise on the gastrointestinal tract. Sports Med 1990;9:159–172.

92 Motil JJ, Ostendorf J, Bricker JT, Klisch WJ: Case report: exercise-induced gastroesophageal reflux in an athletic child. J Pediatr Gastroenterol Nutr 1987;6:989–991.
93 Murray R: The effect of consuming carbohydrate-electrolyte beverages on gastric emptying and fluid absorption during and following exercise. Sports Med 1987;4:322–351.
94 Neufer PD, Young AJ, Sawka MN: Gastric emptying during exercise: effects of heat stress and hypohydration. Eur J Appl Physiol 1989;58:433–439.
95 Nurko S, Dunn BM, Rattan S: Peptide histidine isoleucine and vasoactive intestinal polypeptide cause relaxation of oppossum internal anal sphincter via two distinct receptor. Gastroenterology 1989;96:403–413.
96 Oektedalen O, Flaten P, Opstad PK: LPP and gastric response to a liquid meal and oral glucose during prolonged severe exercise, caloric deficit and sleep deprivation. Scand J Gastroenterol 1982;19:619–624.
97 Oektedalen O, Opstad PK, Schaffalitzky de Muckadell OB: The plasma concentrations of secretin and vasoactive intestinal polypeptide after long-term, strenuous exercise. Eur J Appl Physiol 1983;52:5–8.
98 Oektedalen O, Guldvog I, Opstad PK, Berstad A, et al: The effect of physical stress on gastric secretion and pancreatic polypeptide levels in man. Scand J Gastroenterol 1984;19:770–778.
99 Ollerenshaw KJ, Norman S, Wilson CG, Hardy JG: Exercise and small intestinal transit. Nucl Med Commun 1987;8:105–110.
100 Papaioanides D, Giotis C, Karaginnis N, Voudouris C: Acute upper gastrointestinal hemorrhage in long distance runners. Ann Intern Med 1984;101:719.
101 Pattison ME, Logan JL, Son Lee, Ogden DA: Exertional heat stroke and acute renal failure in a young woman. Am J Kidney Dis 1988;2:184–187.
102 Peat CB, Snijder SH: Opiate receptors: demonstration in nervous tissue. Science 1973;197:1367–1369.
103 Peters O, Peters P, Clarys JT, De Meirleir K, Davis G: Esophageal motility and exercise. Gastroenterology 1988;94:A351.
104 Poortmans JR: Postexercise proteinuria in humans. Facts and mechanisms. JAMA 1985;253:236–240.
105 Porter AMW: Marathon running and the cecal slap syndrome. Br J Sports Med 1982;16:178.
106 Porter AMW: Do some marathon runners bleed into the gut? Br Med J 1983;287:1427.
107 Pruett TL, Wilkins ME, Gamble WG: Cecal volvulus: a different twist for the serious runner. N Engl J Med 1985;312:1262–1263.
108 Qamar M, Reed A: Effects of exercise on mesenteric blood flow in man. Gut 1987;28:583–587.
109 Rattan S, Said SI, Goyal RK: Effect of vasoactive intestinal polypeptide on the lower esophageal sphincter pressure. Proc Soc Exp Biol Med 1977;155:40–43.
110 Rehrer NJ, v Kemenade MC, Meester TA, et al: Nutrition in relation to GI complaints in triathletes. Med Sci Sports Exerc 1990;22:abstr S107.
111 Rehrer NJ: Limits to fluid availability during exercise; PhD thesis, University of Limburg, Maastricht, The Netherlands (1990).

112 Riddoch C, Trinick TR: Gastrointestinal disturbances in marathon runners. Br J Sports Med 1988;22:71–74.
113 Riddoch CJ: Exercise-induced gastrointestinal symptoms, hormonal involvement; PhD thesis, Queens University of Belfast, Ireland (1990).
114 Robertson JD, Maughan RJ, Davidson RJL: Faecal blood loss in response to exercise. Br Med J 1987;295:303–305.
115 Rowell LR, Blackmon JR, Bruce RA: Indocyanine green clearance and estimated blood flow during mild to maximal exercise an upright man. J Clin Invest 1964;43:1677–1690.
116 Ruwart MJ, Rush BD: Prostacyclin inhibits gastric emptying and small-intestinal transit in rats and dogs. Gastroenterology 1984;87:392–395.
117 Said SI: Vasoactive intestinal polypeptide: current status. Peptides 1984;5:143–150.
118 Sandell RC, Pascoe MD, Noakes TD: Factors associated with collapse during and after ultramarathon footraces: a preliminary study. Physician Sports Med 1988;16:86–94.
119 Sanders KM: Role of prostaglandins in regulating gastric motility. Am J Physiol 1984;247:G117–G126.
120 Schaub N, Spichtin HP, Stalder GA: Ischämische Kolitis als Ursache einer Darmblutung bei Marathonlauf? Schweiz Wochenschr 1985;115:13:454–457.
121 Schiff HB, MacSearraigh ETM, Kallmeyer JC: Myoglobinuria, rhabdomyolysis and marathon running. Q J Med 1978;47:463–472.
122 Schofield PM, Bennett DH, Whorwell PJ, et al: Exertional gastro-oesophageal reflux: a mechanism forms symptoms in patients with angina pectoris and normal coronary angiograms. Br Med J 1987;294:1459–1461.
123 Schrier RW, Henderson HS, Tischer CC, Tannen RL: Nephropathy associated with heat stress and exercise. Ann Intern Med 1967;356–376.
124 Schrier RW, Hano J, Keller HI, et al: Renal, metabolic and circulatory responses to heat and exercise. Ann Intern Med 1970;73:213–223.
125 Schwartz A, Vanagunas A, Kamel P: The aetiology of gastrointestinal bleeding in runners: a prospective endoscopic appraisal. Gastrointest Endosc 1989;35:194.
126 Scobie BA: Correspondence. NZ Fed Sports Med 1978;6:31.
127 Scobie BA: Recurrent gut bleeding in five long distance runners. NZ Med J 1985;98:966.
128 Scobie BA: Digestive tract hazards of marathon runners. NZ Med J 1987;100:26.
129 Selby G, Fram D, Eichner ER: Effort-related gastrointestinal blood loss in distance runners during a competitive season. Am Coll Sports Med 1988;20:579.
130 Shibolet S, Lancaster MC, Danon Y: Heatstroke: a review. Aviat Space Environ Med 1976;47:280–301.
131 Sirinek KR, O'Dorisio TM, Gaskill HV, Levine BA: Chronic renal failure: effect of hemodialysis on gastrointestinal hormones. Am J Surg 1984;148:732–735.
132 Sjövall H, Redfors S, Hallback DA, Eklund S, Jodal M, Lundgren O: The effect of splanchnic nerve stimulation on blood flow distribution, villous tissue osmolality and fluid and electrolyte transport in the small intestine of the cat. Acta Physiol Scand 1983;117:359–365.
133 Sörensen JB, Ranek L: Exertional heatstroke: survival in spite of severe hypoglycemia, liver and kidney damage. J Sports Med 1988;28:108–110.

134 Sullivan SN: The gastrointestinal symptoms of running. N Engl J Med 1981;304: 915.
135 Sullivan SN, Champion MC, Christofides ND, Adrian TE, Bloom SR: Gastrointestinal regulatory peptide responses in long-distance runners. Phys Sportsmed 1984; 12:77–82.
136 Stanford B: A stitch in the side. Physician Sports Med 1985;13:187.
137 Stanghellini V, Malagelade JR, Zinsmeister AR, Go VLW, et al: Stress-induced gastroduodenal motor disturbances in humans: possible humoral mechanisms. Gastroenterology 1983;85:83–91.
138 Stanghellini V, Malagelade JR: Digestive motility and beta-endorphin plasma levels in man, under the effect of stress. Ital J Gastroenterol 1983;15:208–220.
139 Stewart JF, Ahlquist DA, McGill DB, Ilstrup DM, Schwartz S, et al: Gastrointestinal blood loss and anemia in runners. Ann Intern Med 1984;100:843–845.
140 Sullivan SN, Lamki L, Corcoran P: Inhibition of gastric emptying by enkephalin analogue. Letter to the Editor, Lancet 1981;ii.
141 Sullivan SN: Exercise-associated symptoms in triathletes. Physician Sports Med 1987;15:105–108.
142 Sullivan SN: The gastrointestinal symptoms of running. N Engl J Med 1981;304: 915.
143 Svanvik J: Mucosal blood circulation and its influence on passive absorption in the small intestine. Acta Physiol Scand 1973;suppl 385.
144 Tache Y: Nature and biological actions of gastrointestinal peptides. Clin Biochem 1984;17:77–81.
145 Thompson DG, Richelson E, Malagelada JR: GI tract, gastric emptying, motility. Perturbation of upper gastrointestinal function by cold stress. Gut 1983;24:277–283.
146 Thompson DG, Richelson E, Malagelada JR: GI tract, gastric emptying and duodenal motility through the central nervous system. Gastroenterology 1982;83:1200–1206.
147 Thompson PD, Funk EJ, Carleton RA, Sturner WQ: Incidence of death during jogging in Rhode Island from 1975–1980. J Am Med Assoc 1982;247:2535–2538.
148 Vandewalle HC, Lacombe JC, Lerievre A, Poirot C: Blood viscosity after 1 h submaximal exercise with and without drinking. Int J Sports Med 1988;9:104–107.
149 Varro GE, Harris JA, Geenen JE: Effect of decreased local circulation on the absorptive capacity of a small intestine loop in the dog. Am J Dig Dis 1965;10:170–177.
150 Vertel RM, Knochel JP: Acute renal failure due to heat injury. Am J Med 1967;43: 435–451.
151 Wade OL, Combes B, Chilos AW, et al: The effect of exercise on the splanchnic blood flow and splanchnic blood volume in normal men. Clin Sci 1956;15:457–463.
152 Wagenmakers AJM, Brouns F, Saris WHM, Halliday D: Maximal oxidation of oral carbohydrates during exercise. Med Sci Sports Exerc 1990;22: abstr S120.
153 Wald A, Van Theil DH, Hoechstetter I, Gavaler JS, Egler KM, et al: Gastrointestinal transit: the effect of the menstrual cycle. Gastroenterology 1981;80:1497–1500.
154 Walsh JH: Gastrointestinal hormones; in Johnson LR, Christensen J, Jackson MJ, Jacobson ED, Walsh JD (eds): Physiology of the Gastrointestinal Tract. New York, Raven Press, 1987, vol 1, pp 181–254.

155 Weisbrot NW, Wiley JN, Overholt BF, Bass P: A relation between gastroduodenal muscle contractions and gastric emptying. Gut 1969;10:543–548.
156 Welbourne RB, Wood SM, Polak JM, Bloom SR: Pancreatic endocrine tumours; in Bloom SR, Polak JM: Gut Hormones. Edinburgh, Churchill Livingstone, 1978.
157 Williams JH, Mager M, Jacobson ED: Relationship of mesenteric blood flow to intestinal absorption of carbohydrates. J Lab Clin Med 1964;63:853–862.
158 Williams RH (ed): Gastrointestinal hormones; in Textbook of Endocrinology. New York, Saunders, 1981, pp 685–715.
159 Woodman D, Hinton J: Catecholamine balance during stress anticipation: an abnormality in maximum security hospital patients. J Psychosom Res 1978;22:477–483.
160 Worme JD, Doubt TJ, Singh A, Ryan CJ, Moses FM, et al: Dietary patterns, gastrointestinal complaints, and nutritional knowledge of recreational triathletes. Am J Clin Nutr 1990;51:690–697.
161 Worobetz LJ, Gerrard DF: Gastrointestinal symptoms during exercise and enduro athletes: prevalence and speculations of the etiology. NZ Med J 1985;98:644–646.
162 Worobetz LJ, Gerrard DF: Effect of moderate exercise on esophageal function in asymptomatic athletes. Am J Gastroenterol 1986;81:1048–1051.

Dr. Fred Brouns, Nutrition Research Center, Department of Human Biology, University of Limburg, PO Box 616, NL–6200 MD Maastricht (The Netherlands)

Exercise, Nutrition and Weight Control

Wim H.M. Saris

Nutrition Research Center, Maastricht, The Netherlands

Introduction

Obesity is one of the most common health problems in affluent societies. In a report of the Dutch Health Council [17], 5% of the adult population was shown to have a body mass index (BMI) of above 30. This index is derived from weight (W) in kilograms and height (H) in meters and is expressed as W/H^2. The J-shaped relationship between mortality and BMI is well documented [41]. There is an exponential increase in mortality above a BMI of 30. The reasons for the increased mortality at low levels of BMI are poorly understood but is beyond the scope of this review. Several diseases can be mentioned in relation to morbidity. Obesity is reported to be associated with the incidence of hypertension, diabetes, and hypercholesterolemia. Corrected for these risk factors, obesity in itself is an independent cardiovascular risk factor [28]. Other diseases mentioned in relation to obesity are several malignancies, e.g. breast and colonic cancer, gallstones and chronic disorders such as osteoarthritis. The psychological consequences are less clear. It has been suggested that obesity affects one's self-image and emotional well-being [50]. However, others could not establish a relationship between emotional disturbance and obesity [54].

Discussing the effectiveness of various intervention strategies, especially the role of exercise, one has to face the question of the benefits of weight reduction in relation to mortality and morbidity. Weight reduction has been shown to benefit several cardiovascular risk factors including hypertension, lipid profile, glycemic control and osteoarthritis [7, 41]. In addition, psychosocial status improves remarkably following weight loss. Therefore, to achieve and to sustain weight reduction to normal levels of BMI of 20–25 is of great value in relation to health status. However, obe-

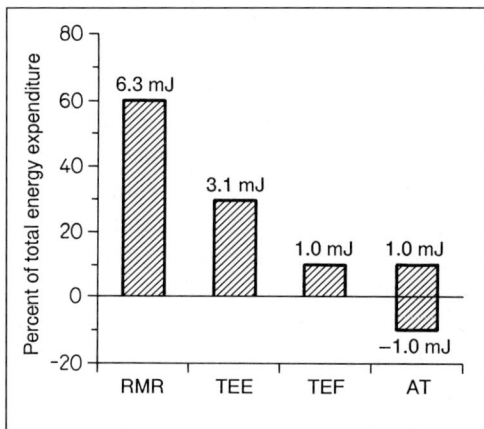

Fig. 1. Compartments of energy expenditure during weight maintenance and the potential modifying effect of adaptive thermogenesis. Approximate percentage of total calories expended by a 70-kg man consuming 10.4 MJ/day [27].

sity is a disease notorious for a high failure rate of treatment. Already in 1959, Stunkard and McLaren-Hume [49] described this gloomy situation as follows: 'Most obese people do not enter treatment for obesity. Of those who do enter, most will not remain. Of those who remain, most will not lose much weight. Of those who lose weight, most will regain it.' Now, 30 years later, one can conclude that, with respect to weight reduction treatment, progress has been made. However, in relation to weight loss maintenance the situation is still hopeless. It is suggested that exercise plays a special role in weight control. It is one of the few factors positively associated with successful long-term body weight maintenance [9].

This review will focus on the role of exercise as one of the components in energy balance and its role in the regulation of metabolism.

Exercise Level and the Regulation of Energy Intake

Management of obesity requires knowledge of body weight regulation. Twenty-four hour energy expenditure can be divided into 4 components as described by Horton [27]. These components are basal metabolic rate (BMR or RMR), thermic effect of exercise (TEE), thermic effect of food (TEF) and adaptive thermogenesis (AT) (fig. 1).

It is plausible to suggest that increasing weight and inactivity may form a vicious circle. As early as 1954, Mayer et al. [33] suggested that the regulation of energy intake was linked with the level of exercise in rats. Energy intake was higher at sedentary levels than at higher levels of physical activity, indicating that regulation of the energy balance is not optimal at lower activity levels. This hypothesis was supported by a study of workers in a jute mill in West Bengal [34]. However, the study had serious defects in both design and analysis [22]. Animal data is supportive of the view that appetite and thus energy intake depend on severity and duration of the exercise and perhaps on sex differences. The interested reader is referred to the classical work of Oscai [27]. This observation is in contrast to the common perception that exercise stimulates appetite leading to an increased food intake even superseding the energy cost of the preceding activities. Although the animal literature is convincing about the nonlinear relationship between energy intake and exercise levels, human studies show equivocal results. Woo et al. [58] did not find an increase in energy intake when comparing 19 days of exercise with 19 sedentary days. In a follow-up study with 3 obese women over 57 days, energy intake increased with exercise but not sufficiently to compensate for the extra expenditure [59]. There is a large variability in daily energy intake and energy expenditure, especially in TEE, indicating that energy balance is best studied in individuals rather than in groups. In a study with British military cadets where intake and expenditure were carefully monitored, no sign of a relationship between energy intake and expenditure, on a daily basis, was observed [18]. The results showed an interesting time lag of 2 days between intake and expenditure. Based on 69 subjects from 6 energy balance studies, Durnin [16] concluded that the results are not supportive of a precise short-term control mechanism. Perhaps the difficulties in measuring habitual daily activities and food intake over a period of time obscure an existing mechanism. If the hypothesis of Mayer is true, one would expect a better relationship between energy intake and expenditure over a shorter period of time at higher activity levels. Recently, Saris [45] observed a close correspondence of energy intake and expenditure on a day-to-day basis in professional cyclists during extreme endurance performance over 21 days. All individual correlation coefficients over the 21-day period were higher than 0.85, indicating that at higher levels of exercise the risk of a deregulation of the energy balance is smaller.

Which mechanisms are possibly involved? At a high rate of energy turnover it is clear that a relative small percentage of deviation leads to

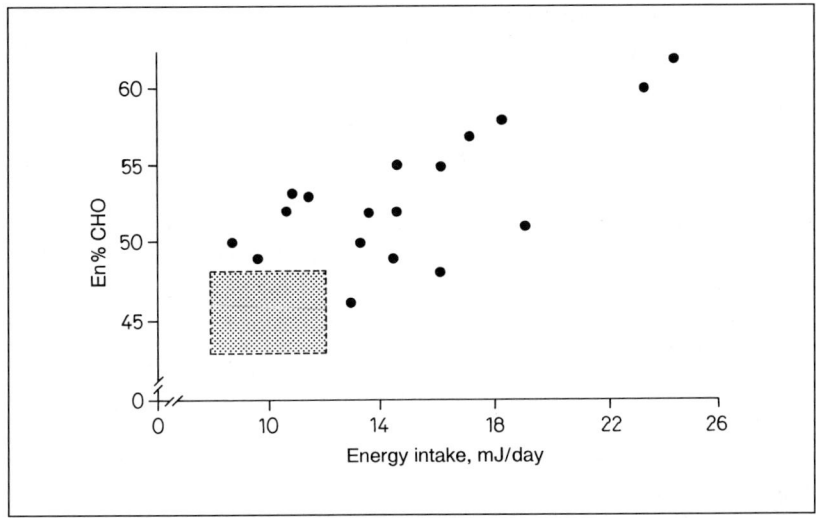

Fig. 2. Relative contribution (En%) of CHO to energy intake in different groups of athletes (n = 232) of both sexes participating in different endurance type sports like cycling, running and swimming. Each point represents a group of athletes (number per group varies between 4 and 50). Shaded area shows range of general intake of Dutch population.

higher absolute levels of energy deficit, or surplus, in contrast to a moderate or a low rate of energy turnover. These larger fluctuations in energy shortage or surplus lead to a stronger regulatory signal in order to keep energy intake in pace with expenditure or to stimulate thermogenesis to get rid of the surplus of calories. In the following paragraphs concerning the different components of energy expenditure, this aspect will be discussed in detail.

Another possible route by which exercise can lead to a better regulation of energy intake is related to the observation that in athletes carbohydrate (CHO) intake is higher at the expense of fat. The need for an enhanced intake of CHO is widely recognized in athletes [10]. An extensive study among elite athletes in The Netherlands showed that these athletes had a higher percentage of CHO in the daily diet compared to the national average intake level of 44% CHO and that this percentage increased with higher levels of energy intake (fig. 2). Why is this observation of interest in relation to the problem of obesity?

There is increasing evidence that the CHO/fat ratio in the diet is of importance in relation to energy balance [48]. The conversion from dietary CHO to fat by the novo-synthesis costs 23% of the original energy from CHO while for dietary fatty acids the cost of deposition as triglyceride is only 3%. Therefore, high fat diets may contribute to obesity through a lower thermogenic effect of food (TEF), as well as higher energy density. Furthermore, the link between CHO intake and CHO balance appears to be much tighter than for fat intake and fat balance [21]. Probably the relatively small CHO storage capacity compared with that of fat has something to do with it. From a regulatory point of view, this difference will lead to priority in keeping CHO balance. Extra CHO intake presumably gives stronger signals to maintain CHO balance. Fat oxidation, which is less than fat intake, favors weight gain. In a study of Tremblay et al. [52] exercise amplified food intake depending on the magnitude of the exercise-induced increase in fat oxidation. This observation is concordant with the notion that energy intake is influenced by the degree to which lipids are oxidized.

It is suggested therefore that improvement of the CHO/fat ratio in an equicaloric diet is one of the factors which can prevent weight gain [32]. The possible relationship between exercise and the improvement of the CHO/fat ratio and thus increased thermogenesis, may therefore be an extra argument to increase activity levels.

Physical Activity and Obesity

Reduced levels of daily physical activity and as a result lower levels of TEE, is often claimed as one of the important factors leading to a disturbance of energy balance. It is the only component which can be increased voluntarily.

Several studies have looked for differences in daily physical activity between obese and lean subjects to explain the positive energy balance. A tendency for obese subjects to be less active have been reported in some studies. However in a number of studies no differences were observed. Of the 16 studies cited in a review concerning this topic [38], 6 reported lower activity levels in obese subjects and 10 failed to find a significant difference.

One of the major problems with respect to these studies is the measurement and definition of daily physical activity. A lower level of movements, for instance, a smaller walking distance during a game, does not automatically mean less TEE. In fact, moving around with a higher body mass implies a higher energy cost. No differences in energy expenditure were found in the studies of Waxman and Stunkard [55] and Saris et al. [43] between obese and lean children, when corrections were made for the larger body mass. The observation scores were, however, significantly lower in the obese ones.

Furthermore, activity is difficult to measure accurately over longer periods without interfering with the daily routine [44]. Perhaps the most promising field technique is the use of stable isotopes. It may bridge the gap between the very accurate but very restrictive methods, like the respiration chamber technique, and the commonly used field techniques, each of which have disadvantages with respect to validity and accuracy. The doubly-labeled water technique provides an overall estimation of energy expenditure over a period of 14 days. Recently, first results were published in which this method was used to measure the energy expenditure levels in obese and nonobese adolescents [3]. No differences were found in TEE expressed as multiples of BMR (1.79 and 1.68 in nonobese/obese males resp. and 1.69 and 1.74 in nonobese/obese females resp.) In general, one can conclude that, although the data are conflicting, there is growing evidence that daily physical activity is not causal to obesity. Once obesity is established, physical activity is generally restricted. This is especially true in the extreme obese state.

Perhaps we have to make one exception to this general conclusion in relation to the smaller movements of the body so-called 'fidgeting'. Ravussin et al. [39] observed a wide variation in fidgeting as measured by radar in the respiration chamber. Evidence was gathered which demonstrated that the extent to which this type of activity performed was a familial characteristic and related to BMI. Although the energy costs of these small movements are dismissed as having only minor effect on the overall energy balance, on the long-term basis, even small contributions will have some impact. Differences in fidgeting may thus contribute to the development of obesity.

Effects of Exercise Alone

Most interesting is the meta-analyses on the effect of training on body weight by Epstein and Wing [20]. Their evaluation was based on 16 studies including 13 with males and 3 with females, lasting from 8 to 26 weeks of training. Thin individuals lost an average of 0.05 kg/week, compared to 0.15 kg/week for more obese subjects. A more recent review by Wilmore [57] showed similar results in 53 studies. This difference in weight loss between lean and obese subjects may be partly explained by the extra cost of exercise due to the overweight of the obese subject. This was demonstrated in an elegant study by Weigle [56], where the experimental subjects wore a vest with weights compensating exactly for body weight loss due to a 700-kcal diet. 24-hour energy expenditure drop with 1.7 MJ in the experimental group, while in the control obese group with identical weight loss a decrease of 3.8 MJ was found. It was concluded that the loss of body weight reduces the energy cost of physical activity sufficiently to account for more than half of the major fall in daily energy expenditure.

Another factor which might play a role of importance is the compensation in daily activities outside the training hours leading to no extra increase in total energy expenditure on a 24-hour basis. We studied this question in two adult groups and a group of 10-year-old obese boys. In lean and obese women, no indication of compensation in normal daily activities was detected, while in lean males, training stimulated physical activity during the nonexercising part of the day [12, 35]. In the 10-year-old obese boys the stimulated effect of the training program on daily physical activity, or other energy-consuming factors such as increased TEF or postexercise increase of BMR, was even greater than in adult subjects. The extra expenditure due to the training was effectively doubled [5].

Looking for an effect of the frequency of the training, Epstein and Wing [20] found no effect between sedentary control groups and training groups which conducted up to three exercise sessions per week (−0.07 kg/week). Average weight loss for those who conducted four training sessions per week was −0.2 kg/week. It is suggested that intensity and duration has a marked effect on changes in body composition based on a shift in fuel preferences during and shortly after exercise. Differences in respiratory exchange ratio (RER) will determine the utilization of CHO and fat stores.

Exercise has been shown consistently to increase fat free mass (FFM) and to decrease body fat (FM). However, this does not automatically lead to changes in body weight. With an intensive type of exercise FFM will increase while FM decreases, resulting in a stable body weight. With less intensive training programs the majority of the studies have shown a significant decrease in FM, sometimes leading to a weight loss [38]. Girandola [24] reported that the rate of loss of FM was more related to duration and intensity than to the total energy expenditure. This decreasing effect of exercise on FM and the preserving effect on FFM reflects lipid mobilization. Higher levels of basal and stimulated lipolysis in suprailiac fat cells have been observed, both denoting an increased fat oxidation capacity during exercise [14]. This leads us to the recently developed hypothesis that overweight had something to do with the capacity to oxidize fat. A reduced capacity automatically leads to an increased lipogenesis. Reduced rates of fat utilization could therefore play a role in the development of obesity.

Metabolic Potential

Of importance to explain this phenomenon of reduced capacity to oxidize fat is the observation of Wade et al. [53], showing that obese men combusted less fat during work at 100 W than did lean men. In the same group he found that the proportion of slow muscle fibers was inversely related to fatness. Slow (type 1) muscle fibers, which are well endowed with mitochondria, usually work oxidatively and use fatty acids as an important fuel source, especially during low and moderate types of exercise. In contrast, fast (type 2) fibers, particularly in those who are not adapted to regular exercise, have fewer mitochondria, and will readily use the glycolytic pathway for energy supply. Indeed, studies in postobese subjects have suggested that reduced fat oxidation may be related to subsequent weight gain [31, 60], and demonstrated that 24-hour RER was correlated with subsequent changes in body weight. Subjects with a higher 24-hour RER, independent from 24-hour energy expenditure, were at 2.5 times higher risk of gaining > 5 kg body weight than those with a lower 24-hour RER. This relationship between muscle fiber type, RER level and weight gain has focussed the attention on the muscle compartment as a possible target organ with a defective thermogenic response. Several studies have recently been conducted to assess the importance of skeletal muscle metabolism as a determinant of metabolic rate.

Astrup et al. [1] demonstrated that about 50% of the increase in oxygen consumption induced by ephedrine may take place in skeletal muscle. Zurlo et al. [60] showed that the differences in muscle metabolism, measured from forearm oxygen uptake, account for part of the variance in BRM and 24-hour energy expenditure among individuals. Astrup et al. [2] also reported that skeletal muscle is the major site of induced thermogenesis in response to CHO feeding via the beta-2-adrenergic receptor stimulation by epinephrine.

Therefore, several factors might influence the variation in muscle energy expenditure among which are fiber type and sympathetic-adrenergic response. Both are effectively influenced by exercise training. Longitudinal and cross-sectional studies with animals and man have demonstrated that dramatic changes can occur in skeletal muscle in response to training [42]. These studies have demonstrated that large increases in the concentration of mitochondrial protein can occur in skeletal muscles during endurance exercise. This increases the capacity of several metabolic pathways including beta-oxidation. The magnitude of this increased oxidative potential varies as a function of the duration and intenstiy of the training program. Furthermore, there is a shift to a greater use of fats during submaximal exercise.

In relation to sympathetic-adrenergic control, Richter et al. [40] showed that endurance training augments the stimulatory effect of epinephrine on oxygen consumption in skeletal muscle. The proposed mechanism responsible for this increased energy expenditure is 'futile cycling' as proposed by Newsholme [36]. A futile cycle is defined as a cycle in which ATP is used to form a phosphorylated product, which is then dephosphorylated with the loss of nonutilizable energy. Newsholme [36] suggested that before, during and after exercise the rate of futile cycling is increased. It is likely that the capacity and responsiveness of substrate cycles to regulatory hormones are increased by exercise training.

Effect of Exercise and Energy Restriction

Evidence is available for both animals and humans that BMR declines up to 20–30% during a energy-restricted diet. In the early classical Minnesota starvation studies of Keys et al. [30], this phenomenon was demonstrated. The largest contribution to this metabolic decline was shown to be due to the loss of FFM. In some studies this decline could be explained completely by the loss of FFM. Other studies, however, showed a larger decline than could be explained on the basis of loss of FFM [11]. Heshka et al. [26] found a disproportionate decrease in BRM from expected values derived from a regression line based on pre-weight-loss data. The decline tended to be greater if the loss of FM was greater. In addition, the decline is said to be limited to the energy-restricted period. However, there is an increasing number of studies showing that postobese subjects have a lower than normal BMR corrected for differences in body composition [13, 23].

Besides this decline in BMR, the TEF is also lower, as a consequence of the lower energy intake. It has been claimed that the TEF decreases more than proportionally during energy restriction, but the results from different studies are equivocal. Both the decrease and increase are probably related to the response to insulin, as insulin resistance depresses TEF [29]. Based on these negative effects of energy restriction on thermogenesis, the combination of a slimming diet and exercise has been suggested as the answer to improved weight and fat loss, preservation of FFM and thus BMR.

A large number of studies have evaluated the combination of both therapies [11]. There is evidence indicating that the combination of energy restriction and exercise leads to an increased loss of FM and that the losses in LBM are less. In a pooled data set of four studies it has been observed that during a 5-week period of a very low calorie diet (VLCD), in a group of 62 obese males and females, a significantly lower weight and FM in the diet-exercise group compared to the diet group (weight: -11.6 and -10.5%, resp., FM: -25.0 and -21.4%, resp.) [46]. BMR decline was the same for both groups, even when corrected for FFM (BMR/kg FFM: -13.4 and -14.6% resp.). There was a marked effect on maximal aerobic power (AP) (V_{O_2}/kg/min: $+18.4$ and $+3.1\%$, resp.). However, the effect of weight loss with or without exercise on AP is not always as clear as was observed in this study. The intensity level of the training program may play an important role in the outcome [25].

Long-Term Effect of Exercise on Weight Maintenance

It is only within the past 10 years that research has been focused on the maintenance of reduced body weight, and in this respect little is known about the effects of continuation of physical activity after weight reduction. Many obese subjects claim that a life-long diet or extreme exercise levels are needed to maintain weight loss. Tremblay et al. [51] demonstrated that ex-obese subjects had to run 90 miles/week to maintain their weight loss and that their body fat was still higher than that of their lean counterparts. These poor long-term results have led to speculations about a postobesity syndrome of a persistently lower energy expenditure. It is of interest to analyze whether BMR corrected for changes in FFM still remained depressed after weight loss.

A persistently lower BMR was observed after 1 week of refeeding compared to pre-diet levels in two studies [4, 6]. Geissler et al. [23] found a 15% lower 24-hour energy expenditure in a group of 16 postobese women, maintaining their weight for at least 6 months, compared with a lean control group. Interestingly, more than 90% of the differences occurred during the day. They concluded that the differences in the postobese are due to a reduced TEF as well as to a reduced BMR. This reduced TEF was also observed in a study of Schutz et al. [47] The authors concluded that the reduced thermogenic response to glucose was not a consequence of obesity, but may be an intrinsic part of the disease which predisposes to weight gain after the treatment of obesity.

In other studies, however, a similar metabolic rate for successful reduced obese and lean subjects has been observed [15, 19]. In the previously mentioned study on the effect of a diet or a diet/exercise program we followed this group over a period of 18 to maximal 42 months posttreatment [13]. Like other investigators, it was observed that the maintenance of body weight after treatment turned out to be very difficult. About 50% of the subjects regained more than 75% of their initial body weight. No differences were found between the diet and diet/exercise groups. In contrast, with these disappointing results, a better weight maintenance was observed in a small number of subjects (13%) who kept exercising after the experimental weight reduction period. These subjects maintained most of their weight loss and had significantly lower body weight and FM at the end of the follow-up period. In addition to the better maintenance of weight loss and body composition, the active group also showed significantly better values with respect to the restoration of BMR (fig. 3).

Remarkably, BMR per kg FFM was still about 10% lower in the unsuccessful group, 18–40 months after finishing the energy restriction period. This marked depression of BMR will have its effect on 24-hour energy expenditure since BMR accounts for about 75% of the total expenditure.

Another factor which may partly explain the differences between the two groups is the effect of exercise on TEF. However, the results from different studies are not convincing. The possible stimulating effect of exercise on TEF is probably not massive enough to explain the differences in long-term results satisfactorily.

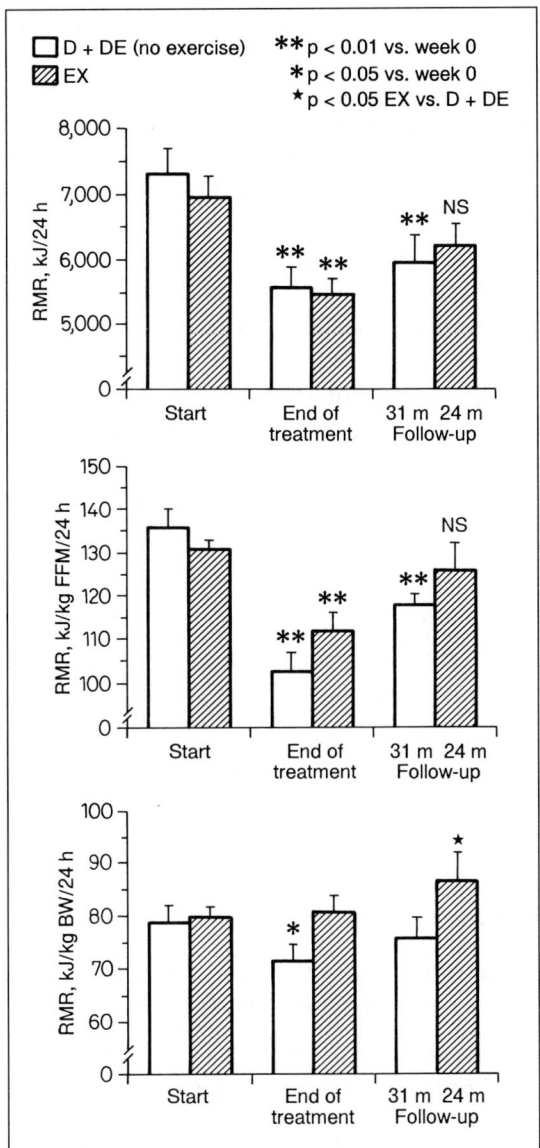

Fig. 3. RMR, RMR per kilogram fat-free mass and RMR per kilogram body weight at the start, after 14 weeks of VLCD treatment with or without exercise (DE and D), and after a follow-up of 31 months for subjects who discontinued exercise (DE + DE, n = 9) and at a follow-up of 24 months for 9 subjects (EX) who maintained exercise activities during the follow-up period [13].

From the presented data it has become clear that exercise plays an important role in the prevention of weight gain. Increasing physical activity seems to be an effective way to increase the metabolic potential in keeping energy balance. Several factors are involved which are not yet fully understood. Nevertheless, all available data indicate that exercise is one of the most powerful effectors in this complex and delicate balance between energy input and output.

Not mentioned in this review are the effects of exercise on the psychological factors. In a successful relapse prevention model, Brownell et al. [8] proposed exercise as one of the most successful tools to prevent relapse. In this respect, more information is needed concerning ways to improve adherence to exercise activities. This is not a unique feature concerning the treatment of obese subjects [8]. They suggested that changing routine activities should result in a better adherence and thus in a better long-term fitness and weight maintenance.

References

1 Astrup A, Bülow J, Madsen J, Christensen NJ: Contribution of BAT and skeletal muscle to thermogenesis induced by ephedrine in man. Am J Physiol 1985;248: E507–E514.
2 Astrup A, Andersen T, Hendriksen O, et al: Impaired glucose-induced thermogenesis in skeletal muscle in obesity. The role of the sympathico-adrenal system. Int J Obes 1987;11:51–67.
3 Bandini LG, Schoeller DA, Diets WH: Energy expenditure in obese and non-obese adolescents. Pediatr Res 1990;27:198–203.
4 Barrows K, Snook JT: Effect of a high protein VLCD on resting metabolism, thyroid hormones and energy expenditure of obese middle-aged women. Am J Clin Nutr 1987;45:391–398.
5 Blaak EE, Westerterp KR, Bar-Or O, Wouters LJM, Saris WHM: Effect of training on total energy expenditure and spontaneous activity in obese boys. Am J Clin Nutr, submitted.
6 De Boer J, van Es AJH, Roovers LA, van Raay JMA, Hautvast JGAJ: Adaptation to low energy intake, studied with whole-body calorimeters. Am J Clin Nutr 1986;44: 585–595.
7 Bray GA (ed): Obesity in America. Proc 2nd Fogarty Int Center Conf on Obesity, No 79. Washington, US DHEW, 1979.
8 Brownell KD, Stunkard AJ, Albaum JM: Evaluation and modification of exercise patterns in the natural environment. Am J Psychiatry 1980;137:1540–1545.
9 Brownell KD, Marlatt GA, Lichtenstein E, Wilson GT: Understanding and preventing relapse. Am Psychol 1986;38:765–782.
10 Coyle EF: Carbohydrate feedings: effects on metabolism, performance and recovery;

in Brouns F, Saris WHM, Newsholme EA (eds): Advances in Nutrition and Sport. Basel, Karger, 1991.
11   Van Dale D, Saris WHM, Schoffelen PMF, ten Hoor F: Does exercise give an additional effect in weight reduction. Int J Obes 1987;22:367–375.
12   Van Dale D, Schoffelen PFM, ten Hoor F, Saris WHM: Effect of adding exericse to energy restriction, 24 h energy expenditure resting metabolic rate and daily physical activity. Eur J Clin Nutr 1989;43:441–451.
13   Van Dale D, Saris WHM, ten Hoor F: Weight maintenance and resting metabolic rate 18–40 months after a diet/exercise treatment. Int J Obes 1990;14:347–459.
14   Depres JP, Bouchard C, Savard R: Levels of physical fitness and adipocyte lipolysis in humans. J Appl Physiol 1984;56:R1157–R1161.
15   Doré C, Miller DS, Shah M: Prediction of energy requirements of obese patients after massive weight loss. Hum Nutr Clin Nutr 1982;36C:41–48.
16   Durnin JVGA: Appetite and the relationship between expenditure and intake of calories in man. J Physiol 1961;156:294–306.
17   Dutch Health Council: Advies inzake Adipositas. Report Health Council, 1984, No 38.
18   Edholm OG, Fletcher JC, Widdowson EM, McCance RA: The energy expenditure and food intake of individual men. Br J Nutr 1955;9:286–300.
19   Elliot DE, Goldberg L, Ruehl SK, Bennet WM: Sustained depression of the resting metabolic rate after massive weight loss. Am J Clin Nutr 1989;49:93–96.
20   Epstein LH, Wing RR: Aerobic exericse and weight. Addict Behav 1980;5:371–488.
21   Flatt JP: Energetics of intermediary metabolism; in Garrow JS, Halliday D (eds): Substrate and Energy Metabolism. London, Libbey, 1985, pp 58–69.
22   Garrow JS: Treat obesity seriously. Edinburgh, Churchill Livingstone, 1981.
23   Geissler CA, Miller DS, Shah M: The daily metabolic rate of the post-obese and the lean. Am J Clin Nutr 1987;45:914–920.
24   Girandola RN: Body composition changes in women: effect a high and low exercise intensity. Arch Phys Med Rehabil 1976;57:297–300.
25   Hagan RD: Benefits of aerobic conditioning and diet for overweight adults. Sport Med 1988;5:144–155.
26   Heshka S, Yang MU, Wang J, Burt P, Si-Sunyer FX: Weight loss and change in resting metabolic rate. Am J Clin Nutr 1990;52:981–986.
27   Horton ES: Introduction: an overview of the assessment and regulation of energy balance in humans. Am J Clin Nutr 1983;38:972–977.
28   Hubert HB, Feinleib M, McNamara PM, Castelli WP: Obesity as an independent risk factor for cardiovascular disease: a 26-year follow-up of participants in the Framingham heart study. Circulation 1983;67:968–977.
29   Jequier E: Energy utilization in human obesity. Ann NY Acad Sci 1987;499:73–82.
30   Keys A, Brozek J, Henschel A, Mickelsen O, Taylor HL: The Biology of Human Starvation. Minneapolis, University of Minnesota Press, 1950.
31   Lean MEJ, James WPT: Metabolic effects of ISO energetic nutrient exchange over 24 h in relation to obesity in women. Int J Obes 1988;12:15–27.
32   Lissner L, Levitsky DA, Strupp BJ, Kalkwarf HJ, Roe DA: Dietary fat and the regulation of energy intake in human subjects. Am J Clin Nutr 1987;46:886–892.

33 Mayer J, Marshall NB, Vitali JJ, Christensen JH, Masheyekki MD, Stare FJ: Exercise food intake and body weight in normal rats and genetically obese adult mice. Am J Physiol 1954;177:544–548.
34 Mayer J, Roy P, Mitra KP: Relation between caloric intake, body weight and physical work: Studies in an industrial male population in West Bengal Am J Clin Nutr 1956;4:164–175.
35 Meyer GAL, Janssen GME, Westerterp KR, Verhoeven F, Saris WHM, ten Hoor F: The effect of a 5-month endurance training program on physical activity: evidence for a sex difference in the metabolic response to exercise. Med Sci Sport Exerc, in press.
36 Newsholm EA: A possible metabolic basis for the control of body weight. N Engl J Med 1980;302:400–405.
37 Oscai LB: The role of exercise in weight control; in Wilmore JH (ed): Exercise and Sport Sciences Reviews. New York, Academic Press, 1973, vol 1, pp 103–123.
38 Pacy PJ, Webster J, Garrow JS: Exercise and obesity. Sport Med 1986;3:89–113.
39 Ravussin E, Lillioja S, Anderson TE, Christin L, Bogardus C: Determinants of 24 h energy expenditure in man. Methods and results using a respiratory chamber. J Clin Invest 1986;78:1568–1578.
40 Richter EA, Christensen NJ, Ploug T, Galbo H: Endurance training augments the stimulatory effect of epinephrine on oxygen consumption in perfused skeletal muscle. Acta Physiol Scand 1984;120:613–615.
41 Royal College of Physicians: Obesity. A report of the Royal College of Physicians. J R Coll Physicians 1983;17.
42 Saltin B, Gollnick PD: Skeletal muscle adaptability. Significance for metabolism and performance; in Peach LD, Adrian RH, Geiger SR (eds): Handbook of Physiology: Skeletal Muscle. Baltimore, Williams & Wilkins, 1983, pp 555–631.
43 Saris WHM, Cramwinckel AB, Elvers JE, Binkhorst RA: How inactive are obese boys (abstract). 4th Int Congr on Obesity, New York, 1983, p 41.
44 Saris WHM: Habitual physical activity in children: Methodology and findings in health and disease. Med Sci Sports Exerc 1986;18:253–263.
45 Saris WHM: Physiological aspects of exercise in weight cycling. Am J Clin Nutr 1989;49:1099–1104.
46 Saris WHM, van Dale D: Effects of exercise during VLCD diet on metabolic rate, composition and aerobic power: pooled data of four studies. Int J Obes 1989;13: S169–S170.
47 Schutz Y, Golay A, Felber JP, Jequier E: Decreased glucose-induced thermogenesis after weight loss in obese subjects: A predisposing factor for relapse of obesity. Am J Clin Nutr 1984;39:380–387.
48 Sims EA, Danforth E: Expenditure and storage of energy in man. J Clin Invest 1987; 79:1019–1025.
49 Stunkard AJ, McLaren-Hume M: The results of treatment for obesity. Arch Intern Med 1959;103:79–85.
50 Stunkard AJ: Obesity. Philadelphia, Saunders, 1980.
51 Tremblay A, Despres J, Bouchard C: Adipose tissue characteristics of ex-obese long distance runners. Int J Obes 1984;8:641–648.
52 Tremblay A, Plourde G, Despres JP, Bouchard C: Impact of dietary fat content and fat oxidation on energy intake in humans. Am J Clin Nutr 1989;49:799–805.

53 Wade AJ, Marbut MM, Round JM: Muscle fiber type and etiology of obesity. Lancet 1990;335:805–808.
54 Wadden TA, Stunkard AJ: Social and psychological consequences of obesity. Ann Intern Med 1985;103:1062–1067.
55 Waxman M, Stunkard AJ: Calorie intake and expenditure of obese boys. J Pediatr 1980;96:187–193.
56 Weigle DS: Contribution of decreased body mass to diminished thermic effect of exercise in reduced-obese men. Int J Obes 1988;12:567–578.
57 Wilmore JH: Body composition in sport and exercise: directions for future research. Med Sci Sports Exerc 1983;15:21–31.
58 Woo R, Garrows JS, Pi-Sunyer FX: Effect of exercise on spontaneous calorie intake in obesity. Am J Clin Nutr 1982;36:470–477.
59 Woo R, Garrow JS, Pi-Sunyer FX: Voluntary food intake during prolonged exercise in obese women. Am J Clin Nutr 1982;36:478–484.
60 Zurlo F, Lillioja S, Puente A, et al: Low ratio of fat to carbohydrate oxidation as predictor of weight gain: Study of 24 h RQ. Am J Physiol 1990;259:E650–E657.

Wim H.M. Saris, MD, Nutrition Research Center, Department of Human Biology, Biomedical Center, NL–6200 MD Maastricht (The Netherlands)

Subject Index

Acetylcholine
  dietary effects on synthesis 101
  neurotransmitter 101
Acidosis
  development
    during exercise 149, 150
    low carbohydrate diet as cause 155, 160
  enzyme sensitivity to pH 150
  fatigue relationship 150
  sodium
    bicarbonate
      exercise performance influence 157, 158
      mechanism of action 159
      reversal of acidosis 156
      side effects 159
    citrate
      exercise performance influence 157, 158
      mechanism of action 159
      reversal of acidosis 156
      side effects 159
Adenosine triphosphate
  consumption during exercise 146–149
  creatine phosphate relationship 147, 148
  energy source for contracting muscles 147

Amino acids
  carnitine effects on branched-chain oxidation 116
  exercise effect
    oxidation 23–25, 30, 114, 115
    plasma concentrations 81, 103–105
  muscle concentration in overtraining syndrome 90–92
  neurotransmitter precursors 80

Branched-chain keto acid dehydrogenase
  activation by exercise 24
  hydroxytryptamine effect on activity 82

Calcium
  channel alteration by free radicals 61, 71
  phospholipase A_2 activation 61, 71
Calorie, *see* Energy expenditure
Carbohydrates, *see also* Glycogen
  absorption with decreased blood flow 177
  conversion to fat 2
  craving
    serotonin role 99, 100
    symptoms 100
    treatment drugs 99, 100
  feeding
    benefits during intermittent exercise 7

Subject Index

Carbohydrates (continued)
    chromium loss inhibition 42
    comparison of metabolic responses during low and high intensity exercise 8, 9, 151–153
    effects during prolonged exercise 4–7, 153
    gastric emptying rate effects 135–138
    intake importance to obesity 203, 204
    mood effects 98
    oxidation rate
        dependence on sugar type 178
        during exercise 1, 12
    precompetition meal composition 3, 4
    storage capacity of body 1, 2

Carnitine
    acetylation
        carnitine acetyltransferase as catalyst 114
        importance of increased free CoA pool 114, 116, 121, 122
    biosynthesis not enhanced by lysine or methionine 123, 124
    branched-chain amino acid oxidation stimulation 116
    D-carnitine
        impurity in L-carnitine preparations 123
        toxicity 123
    dietary sources 110
    excretion 111, 118, 119
    exercise effects on concentrations
        muscle 117, 118
        plasma 118
        urine 118, 119
    fatty acid oxidation enhancement 112, 113, 121
    modified forms 111
    sites of concentration 111, 117
    supplementation effects
        athletic performance
            aerobic 120
            anaerobic 120, 122
        benefits for athletes 113, 114
        blood flow 122
        concentrations in athletes 119, 120
        dosage recommendations 124
        fatty acid oxidation 121
        heart rate 120, 121, 123
    synthesis from trimethyllysine 111
    water solubility 124

L-Carnitine, see Carnitine

Carnosine
    antioxidant 67
    distribution 68

Catalase
    activity levels 66
    free radical scavenger 66

Ceruplasmin
    copper as cofactor 45, 46
    iron transport importance 48
    levels increased after exercise 48

Choline
    acetylcholine synthesis effect 101
    exercise effects on plasma level 104, 105

Chromium
    absorption efficiency 39
    conversion to usable form 39
    deficiency
        stress factors 39
        symptoms 39
    dietary sources 40
    exercise effects
        serum concentration 40, 41
        sweat losses 43
        urine excretion 41, 42
    glucose tolerance relationship 40
    insulin activity effects 39, 44
    recommended intake 39
    supplementation effects 44, 53
    toxicity 40

Copper
    athletic performance effect 49
    deficiency
        stress factors 45, 46
        symptoms 45, 46
    dietary sources 46
    enzyme cofactor 45
    exercise effects
        excretion rate 47
        serum level 46, 47
        sweat losses 47, 48
    recommended intake 45

Subject Index

Creatine phosphate, consumption during exercise 146–149
Cytochrome oxidase, copper as cofactor 45

Dehydration, *see also* Fluids
  effect on blood flow 175
  supplement recommendations 142, 143
  sweating as cause 128
Dexfenfluramine, effects on serotonin 99, 100
Diarrhea
  incidence
    runners 167–169
    triathletes 170–172
  prostaglandin role 188
Dopamine, release modulated by tyrosine 102

Electrolytes, gastric emptying rate effects 138
Electron paramagnetic resonance, free radical measurement technique 62–63
Emotional stress
  catecholamine induction 188
  gastric emptying rate decrease 188
Endurance sports, recommended dietary percentage
  carbohydrates 2
  fat 2
  protein 2
Energy expenditure
  during
    endurance sports 2
    intense training 2
    Tour de France 9, 10
  metabolic sources during prolonged exercise 6

Fasting, effect on
  glycogen concentration 154
  high-energy exercise 153
Fat
  carnitine-mediated transport 112, 113
  oxidation rate during exercise 1
Fatigue
  ammonia contribution 10
  delay by carbohydrate feeding 5–8

Fluids
  absorption 129
  gastric emptying rate
    beverage temperature effects 138, 139
    carbohydrate concentration 135–138, 187
    dehydration effects 139
    electrolyte effects 138
    emotional strain effects 188
    exercise effects 187, 188
    experimental difficulties of study 130
    factors influencing 129, 132, 142
    gastric volume stimulation 134, 135
    gastrointestinal energy supply 142
    osmolality effects 133
    significance to water and nutrient availability 139–142
    solution composition effects 130–132, 134
    training status importance 132
  rehydration, suggested supplements 141–143
  supplements, ingestion considerations 129
Fluvoxetine, effects on serotonin 100
Free radicals, *see also* Lipid peroxidation
  cellular antioxidant defenses 65
  environmental sources 62
  measurement of activity 62–65
  mitochondrial production 59, 60
  neutrophil formation 62
  sites of generation 60
  targets of damage 60, 61
Fructose, conversion to glucose in liver 2

Gastric emptying, *see* Fluids
Gastrin, exercise effects on level 181, 182, 184, 185
Gastroenteropancreatic hormones
  exercise effects on levels 181–185
  function 181
Gastrointestinal disorders
  exercise effects on gastrointestinal blood flow 174–176
  factors controlling intestinal blood flow 175, 176

Gastrointestinal disorders (continued)
  gastroenteropancreatic hormone role 181–185
  gut motility
    defecation role 189
    exercise effects 189, 190
    training effects 190
  incidence
    runners 168–170
    triathletes 170–172
  intestinal bleeding
    causes 177, 180
    incidence
      cyclists 180
      runners 179, 180
      skiers 180
    testing 179
  physiological functions responsible 174
  prevention measures 191
  scope of problem in athletes 166, 167
  symptom classification 173
Glucagon, exercise effects on level 181, 182
Glucose
  fatigue reversal by infusion 7
  plasma concentration decrease during prolonged exercise 4, 5
  rate of metabolism by immune system 86
Glutamine
  exercise effects on plasma concentrations 89, 90
  immune system
    rate of metabolism by cells 86, 87
    role 85–87, 91, 92
  production and release by muscle 87
  transport 88
Glutathione
  antioxidant 67
  distribution 67
  peroxidase
    activity levels 66
    free radical scavenger 66
Glycogen
  chromium effects on synthesis 43, 44
  maximizing prior to competition 3, 12

  muscle
    concentration at rest 154
    resynthesis
      rate 10
      dietary maximization following exercise 10, 11
      during exercise 11, 154, 155
    stores during intense exercise 151, 152, 155

Heartburn, causes in athletes 185, 186
Hydroxytryptamine
  brain functions
    autonomic function 84
    endocrine function 84
    motor neuron excitability 84
    sleep 83
  exercise effect on concentration 81, 82
  neurotransmitter 80
  peripheral nerve function 84
  tryptophan as precursor 80

Immune system
  exercise effects
    high intensity adverse effects 85
    low intensity benefits 85
  glutamine role in immunosuppression 85–87, 91, 92
  nutrients used by cells 86

Kidney
  dysfunction in athletes 186
  trauma from sports 186, 187

Lactic acid, *see also* Acidosis
  carbohydrate diet effects on stores 151, 152
  dehydrogenase, zinc as cofactor 38
  formation rapid upon high intensity exercise 148, 149
  muscle buffering of pH 149, 150
Lean body mass, high-protein diet effect 29
Lecithin, effect on acetylcholine synthesis 101
Leucine oxidation, exercise effect 23, 24

Subject Index

Lipid peroxidation, products 63–65
Lysyl oxidase, copper as cofactor 45

Maximum oxygen intake, definition 147
Methylhistidine
    found exclusively in contractile
        proteins 25
    protein degradation marker 25, 26
    strength exercise effect on urinary level
        30, 31
Monoamine oxidase, copper as cofactor
    45
Motilin, exercise effects on level 181, 182

Nausea, incidence
    runners 168, 169
    triathletes 170–172
Neurotensin, exercise effects on level 182
Neurotransmitters, nutrients affecting
    activity 94
Nitrogen balance
    during
        moderate exercise 18–21
        prolonged exercise 18
        sedentary state 18, 20
    measurement difficulty in humans 17, 18
    relationship to protein intake 20
Norepinephrine, depressed by tyrosine
    supplementation 102

Obesity
    carbohydrate intake importance 203,
        204
    diet and exercise control 208, 209
    exercise effects
        long term 209–212
        short term 205, 206
    fat oxidation potential in subjects 207,
        208
    mortality at high body mass index 200
    physical activity relationship 204, 205
    risk factors 200
    treatment difficult 201
Overtraining syndrome
    definition 79
    glutamine deficiency observed 90–92
    hydroxytryptamine role 83, 84
    imbalanced amino acid hypothesis
        80–84
    signs and symptoms 83

Pancreatic polypeptide
    exercise effects on level 181–183
    gut motility role 190
Phenylalanine
    free radical sensitivity 61
    protein degradation marker 25, 26
Phospholipase A_2
    activation by calcium 61, 71
    vitamin E effects 71, 72
Protein, *see also* Nitrogen balance
    degradation
        exercise effects 25, 26, 30, 31
        markers 25, 26
    dietary intake
        daily requirements for athletes 31, 32
        during exercise 18–21
        high-protein diet effects 29
        nitrogen balance relationship 20, 27
        serotonin level effects 95–97
    gender differences in utilization 21, 22
    historical overview of importance in
        diet 15–17
    intake of strength athletes 26–28
    metabolism schematic 23

Reflux, *see* Heartburn

Secretin, exercise effects on level 182
Serotonin
    drugs affecting activity 99, 100
    exercise effects on level 103
    physiological range of release over time
        97
    protein intake effects on levels 95–97
    role in
        carbohydrate craving 99
        nutrient choice 99
        SADS 99
    synthesized from tryptophan 95
Sodium
    bicarbonate, *see* Acidosis
    citrate, *see* Acidosis

Subject Index

Somatostatin, exercise effects on level 185
Sports anemia, prevention by dietary protein 18, 19
Strength exercise
 amino acid oxidation 30
 nitrogen balance studies 26–29
 relationship between protein intake and nitrogen balance 27, 28
Strong ion difference
 definition 156
 dietary factors affecting 156
Superoxide dismutase
 activity levels 66
 copper as cofactor 45
 free radical scavenger 66
Swimming, trace element losses 52

Trace elements, *see* Chromium, Copper, Zinc
Tryptophan
 dietary supplementation dangers 98
 exercise effect on concentration 81, 82
 hydroxytryptamine precursor 80
 inverse relationship between plasma and brain concentrations 95–97
 serotonin effects with diet 95–97
Tyrosine
 blood pressure effects 102
 dopamine release modulator 102
 exercise effects on plasma level 103
 norepinephrine depression after supplementation 102, 103
 protein degradation marker 25, 26

Urea
 serum increase during exercise 21, 22
 sweat excretion 21
Urination, incidence in runners 170, 172

Vasoactive intestinal peptide
 exercise effects on level 181–184
 gut motility role 190
Vitamin
 A
  antioxidant 67
  distribution 67

C
 antioxidant 67
 distribution 67
E
 antioxidant 67
 distribution 67
 exercise effects with diet alterations
  deficient 70, 71
  sufficient 69, 73
  supplemented 71–75
 membrane effects 72
 skeletal muscle importance 68, 69
Vomiting, incidence
 runners 168, 169
 triathletes 170–172

Weight control, *see also* Obesity
 carbohydrate intake importance 203–204
 components of regulation
  adaptive thermogenesis 201
  basal metabolic rate 201
  thermic effects of
   exercise 201
   food 201
 energy intake as a function of activity 202
 exercise effects
  long term 209–212
  short term 205, 206
 lifting, *see* Strength exercise

Zinc
 deficiency
  recommended intake 49
  symptoms 49
 dietary sources 49, 50
 distribution in body 49
 effect on exercise performance 53
 enzyme cofactor 38, 49
 exercise effects on
  excretion rate 51
  serum level 50–52
  sweat loss 52
 supplementation 53
Zymelidine, effects on serotonin 100